PHYSICS IN CRISIS

From Multiverses to Fake News

PHYSICS IN CRISIS

From Multiverses to Fake News

Bruno Mansoulié
Université Paris-Saclay, CEA, France

Translated from the French by Nanette McGuinness

 **World Scientific**

NEW JERSEY · LONDON · SINGAPORE · BEIJING · SHANGHAI · HONG KONG · TAIPEI · CHENNAI · TOKYO

Published by

World Scientific Publishing Europe Ltd.

57 Shelton Street, Covent Garden, London WC2H 9HE

Head office: 5 Toh Tuck Link, Singapore 596224

USA office: 27 Warren Street, Suite 401-402, Hackensack, NJ 07601

Library of Congress Cataloging-in-Publication Data

Names: Mansoulié, Bruno, author.

Title: Physics in crisis : from multiverses to fake news / Bruno Mansoulié,
    Université Paris-Saclay, CEA, France ; translated from the French by Nanette McGuinness.

Other titles: Physique en crises. English

Description: New Jersey : World Scientific, [2023] | Includes index.

Identifiers: LCCN 2022006839 | ISBN 9781800612341 (hardcover) |
    ISBN 9781800612358 (ebook for institutions) | ISBN 9781800612365 (ebook for individuals)

Subjects: LCSH: Physics--Philosophy.

Classification: LCC QC6 .M329513 2023 | DDC 530.01--dc23/eng20220422

LC record available at https://lccn.loc.gov/2022006839

British Library Cataloguing-in-Publication Data

A catalogue record for this book is available from the British Library.

For any available supplementary material, please visit
https://www.worldscientific.com/worldscibooks/10.1142/Q0364#t=suppl

Desk Editors: Jayanthi Muthuswamy/Adam Binnie/Shi Ying Koe

Typeset by Stallion Press
Email: enquiries@stallionpress.com

Printed in Singapore

Preface

S cience and technology have never been a larger part of daily life than they are today. All over the planet, even in the most remote areas, people carry smartphones — incredible amalgamations of the latest scientific advances that incorporate electronics (which are based on Quantum Theory) and GPS systems (which communicate with satellites and contain a correction from General Relativity). Technology has, indeed, become ubiquitous in modern society. However, this amazing effectiveness cannot hide the fact that science, particularly physics, is facing three major crises.

The first crisis affects all of science and is social in nature. The problem? The gap between science and the public has never been as wide as it is now. At a minimum, people take science for granted and pay little attention to the progress made or questions asked. Or people are concerned about its negative implications, rightfully questioning the risks and impact of technology on their way of life. At worst, people reject science and embrace all kinds of beliefs and fake facts, such as the flat Earth hypothesis, to only name one.

The second is specific to fundamental physics. We have long known that our understanding of physics is flawed, or at least incomplete. Observations of the cosmos at very large distances show that something exists (so-called dark matter and dark energy) — something not described by our current theories. In terms of pure theory, we are facing several major problems, of which the most blatant is that our two most foundational theories — Quantum Theory and General Relativity — are incompatible.

Despite this incompatibility, physics remains an effective means of exploring our environment and the universe from the smallest scale (particles) to the largest (cosmology). Still, these problems have now been with us for quite a while; numerous attempts to solve them (such as the well-known "Superstring Theory") have failed, and experiments have not yielded any clues toward a possible solution.

Are we, therefore, stuck with this impasse?

Perhaps — or perhaps not. One recent development — and also the third crisis — is neither an advance in theory nor an unexpected observation: rather it is the rise of artificial intelligence, in particular, machine learning and big data. AI is already a game-changer in all of our lives, and it could soon be one for the way we do science as well. This could lead to a serious crisis if it forces us to abandon our cherished theories for "black box" algorithms.

Yet the seeds of a solution to the first two crises may lie in this third one: AI could also lead to an answer — or part of one. As people become more and more accustomed to algorithms and their not-so-obscure operations, the gap between the public and science may well narrow. Maybe the public will even find AI easier to comprehend than Quantum Theory and General Relativity!

As for the internal crisis in physics, AI algorithms may provide the missing link between our theories and observations. This link would not necessarily be a large "Aha!" advance, such as those made by Newton or Einstein. But who says that we will always solve the problems in physics with an "Aha" theory?

Before we address such deep, thorny questions, it might be good to start with much simpler ones: how do we represent the world around us? What is a theory in physics, and further, what is a *good* theory?

Using simple language and insightful examples, *Physics in Crisis* begins with these basic questions and then turns to the more complex ones. The author ends with his own, personal viewpoint, a call to: reinstate the values of Enlightenment; take advantage of these crises to reconcile the public with science; foster a renewed appreciation of the beauty of science; and espouse a rationalist spirituality!

About the Author

Bruno Mansoulié is a senior scientist at the Particle Physics Division of the Commissariat à l'Energie Atomique et aux Energies Alternatives, CEA (Alternative Energies and Atomic Energy Commission in France), which he has led for five years. He contributed to the hunt and discovery of the Higgs boson at the Large Hadron Collider (LHC), the world's largest particle collider, which is operated by/at CERN (the European Laboratory for Particle Physics). He enjoys reaching out to science enthusiasts during conferences and through various popular media.

About the Translator

Award-winning opera singer **Nanette McGuinness** is the translator of 80 books and graphic novels for adults and children from French, Italian, and German into English. Her translation of *Luisa: Now and Then* was a 2019 Stonewall Honor Book.

Photo by Russ Fischella

Contents

Chapter 1

Physics: At the Heart of the Crisis in Fundamental Science

Society in Crisis

Physics is considered a "hard" science: its objects are inanimate and, in principle, we can reproduce its experiments and measurements at will. Robust, rigorous, deductive mathematical procedures support its theories. As physics does not deal with living beings, it does not give rise to ethical or social questions, or at least not as immediately as in medicine or biology. Ever since Galileo, religion no longer opposes it directly, outside of a few zealots. Even American creationists, who dispute the theory of evolution, don't challenge physics, itself — at least not publicly. In everyday society, far from researchers and labs, men and women throughout the entire world use sophisticated objects that come from an understanding of physics: a cell phone is a fine symbol of the incredible sum of scientific knowledge (especially from the hard sciences, and particularly physics) at everyone's disposal.

You might think, then, that physics had reached the ideal position for a science: of having an established, well-understood, well-accepted body of knowledge, and with this understanding of nature as a starting point, being in a position to offer numerous applications that would be accessible to the greatest number of people. The field would, therefore, quietly carry on, moving along its path from discovery to discovery, incorporating new information into the current body of knowledge, and making new items, methods, and possibilities available to human beings. Bit by bit, the new concepts would percolate from labs to college classes, then to high schools and elementary schools, smoothly spreading through society. With a little time, the most modern ideas would replace old ones. Whether to use this or that new discovery would be the subject of peaceful discussion between citizens who, even if they certainly did not know as much as researchers, would share enough of an understanding of the basics to be able to make enlightened decisions.

I see you smile: we are very, very far from that. Society misunderstands and very often, mistrusts science. Long years of education do not seem to close the gap between scientists and the rest of the population, and this even for "hard" sciences like physics. It makes perfect sense that it might be difficult to connect scientific knowledge to what everyone feels in their own bodies. But in the inanimate science of physics, these types of issues do not exist, and nothing should get in the way of sharing knowledge with everyone. However, even though the scientific community (researchers, engineers, professors) agrees with the advances in modern physics, it remains mysterious to the rest of the world.

Everyone has a smartphone, but very few people have any idea of the concepts underlying its electronics. In this case, we can always make the excuse that the science is relatively recent. Quantum Theory,[1] which is the basic theory that allows us to understand and design electronics, dates to the beginning of the 20th century; its basic, tangible component, the transistor, was invented in 1947. Thus, it took more than a century for these scientific ideas to spread and become assimilated. It took so long because the subject matter is so complex: everyone agrees that Quantum Theory is hard to teach and even harder to popularize. But what about the oldest, simplest concepts? In Western countries, every survey that asks whether the Earth revolves around the sun or the sun revolves around the Earth gets a certain percentage of respondents — a minority, of course, but a significant number — who favor the latter. And that's despite the large stream of information that constantly demonstrates scientists' mastery of the mechanics of celestial bodies. The same person who watches the Philae space probe land on the Churyumov–Gerasimenko comet on TV can answer in a survey that the sun goes around the Earth.

So what is the origin of the gulf between the public and science — and, in terms of my focus, physics?

We could blame the media. As with other subjects, media attention lends itself poorly to in-depth examination and concentration. Controversy often replaces reflection: it's impossible on a TV show to calmly tackle scientific subjects that are important for society,

---

[1]"Quantum Theory," the theoretical body of knowledge commonly used by physicists, is also often called, "Quantum Mechanics," a historical term that is too limited now. In addition, the adjective, "quantum," without a capital letter, can refer to more general aspects.

such as nuclear energy. To be sure, the discussion is somewhat subjective, as everyone can have their own opinion about the best balance between potential risk and hoped-for comfort. But a large part of the argument should be scientific in nature, using a hard, all-purpose science — physics. Why, then, aren't these arguments the subject of broad consensus, which would allow for a calm discussion? It's too easy to blame the media as the source of this huge gap in understanding. The media just gives viewers what they want to see. It honestly reflects the public's desire for science and this desire for science does, indeed, exist. Admittedly, science isn't a priority compared to subjects such as politics, the economy, or safety — which have an immediate impact on society — or sports — where uncertainty keeps viewers at the edge of their seats — but it still has a regular place in the media.

In my own field of elementary particle physics, the media thoroughly covers events such as the commissioning of CERN's large accelerator (LHC) in Geneva in 2008. We were able to share this scientific adventure with the public: we had the opportunity to talk about the years spent designing and building this gigantic instrument; we described international collaborations between thousands of researchers and engineers. For this occasion, newspapers, TV stations, and radio stations let us take center stage. And they did so again in 2012, when we announced the discovery of the Higgs boson via the accelerator and associated experiments. Many physicists, each in their own way, stepped into the spotlight to describe this new, intriguing particle and all the concepts that its experimental existence proves.

So the public's appetite is not at issue either, at least not in our society and for those whose basic needs are being adequately met. Our talks to the general public are packed. Of course, these are amateurs who follow the subject, but even when I gave a lecture about the Higgs boson to the general public in a shopping center, I still had an audience. And when I meet someone and tell them about my work, it's always a pleasure to see that it fires up their imagination. Even when I describe the size of the instruments — and by implication, their cost — it is rare that someone says the money could be used for other, more "useful" things.

We could blame education, and some do so. Years of classes in the natural sciences — and physics, for some — don't manage to transmit the knowledge or even the concepts that our technological society is based upon, or at the very least, they do so poorly.

The problem is profound, at least in France. It seems as though the physics studied in school is almost completely disconnected from the image that students have of the world. On one side are universal scientific theories; on the other, a personal, sense-based perception of the world — with no connection between the two. Moving beyond mere criticism, attempts are being made to remedy this situation, such as the French program, "La main à la pâte" (Hands-On Learning) begun by the Nobel Prize winner Georges Charpak, who suggests that children discover scientific reasoning on their own and create their own image of the world for themselves. Today, the program reaches thousands of students — but that barely scratches the surface. The extent of its impact remains to be seen, for example, on how scientific questions are treated by the media, and in social media, too.

Under these circumstances, it's no wonder that a feeling of being puzzled or even excluded by science turns into discomfort and then distrust. Despite indisputable advances using information based on scientific evidence, all kinds of beliefs thrive, including toward hard sciences such as physics. We reach the peak of this paradox when screens that are connected — pure products of scientific advances, the language for which (the "web") was invented at that temple of physics, CERN — serve as a vehicle for the most ill-founded, least comprehensive claims, ones least deserving of any consensus.

When the LHC was commissioned, a rumor appeared that the new accelerator could create a black hole that would quickly devour the area, then Geneva and the lake, and finally the entire Earth. I received several letters begging me to stop it, including a very sincere one handwritten in Russian.

CERN managed this totally unfounded rumor well, and the LHC got off the ground. But the takeaway from this touching letter stuck with me: would we scientists be so mad as to put the existence of our planet at risk simply for the pleasure of our research? How, then, are we seen? Aren't we decent citizens, with families and friends?

The breach between science and society goes beyond physics. And yet, once again, one might have thought that the "hardest" sciences should also be the ones most readily based on consensus and the easiest to disseminate. This breach isn't recent, even though the tremendous scientific developments of the 19th and 20th centuries, especially in physics, have brought concrete evidence of its effectiveness. During this time, physics has repeatedly turned experts' notions of the world upside down. But for non-experts, science hasn't really changed the vision that each of us has of reality. For non-experts, the world's modern scientific vision is restricted to fantasy, in science fiction, for example, with space travel and time paradoxes. But the everyday world has remained quite "classical," without Quantum Theory or relativity having had any impact. The non-experts have been left to the side, as far as concepts are concerned. Hopefully, they have seen the tangible advances and benefited from all the new tools. Or are they perhaps simply afraid of the future?

The serious imbalance between scientific advances and the global perception of these advances is quite old, but until now, forward progress has more or less managed to hide this imbalance. After all, it doesn't really matter whether concepts are understood and integrated if their applications satisfy us. Today, the question is whether this dynamic is still possible; can advances in physics continue to feed our hunger for novelty and innovation if they're only shared at a surface level? And most importantly, does it make sense to continue this dissonance?

The Crisis in Physics

The crisis that I would now like to address is clearly of an entirely different nature. It's an internal crisis, known only to professionals and amateurs who follow developments in so-called "fundamental" physics, which tries to describe elementary particles and forces and the universe as a whole. "Fundamental" is used here without value judgment: it's about describing the "basic principles," the "elements" of the world, and how they construct the universe, without claiming that this area of physics has some precedence over other areas in physics or other sciences.

I will return to the different aspects of this crisis, but first let's quickly try to describe the current situation. Fundamental physics follows two major paths: describing the basic components of the world (toward the infinitely small) and delineating the different structures in the universe (toward the infinitely large).

This first has led to the discovery of the increasingly tiny building blocks of matter — atoms and nuclei — all the way to the particles we consider elementary today — quarks, electrons, neutrinos, etc.

The second has allowed us to understand the solar system, the sun itself, the other stars, galaxies, and galaxy clusters, and focuses today on the entire universe.

With regard to the "infinitely small," today we have a well-defined notion of elementary particles and their interactions supported by a set of unambiguous mathematical laws. The underlying theory is based on two cornerstones: Special Relativity and Quantum Theory, two major concepts discovered at the very start of the 20th century.

All experimental measurements performed in labs agree with this theory. The measurements are made under completely different conditions, for example, with particle accelerators of greatly differing types and energies. The theory's success, and the fact that no lab measurements have ever contradicted it to date, has earned it the name since the 1980s of the "Standard Model." It's important to stress the extraordinary ability of the Standard Model to predict numerical values for certain physical quantities: for example, it predicts the "magnetic moment" of an electron with an accuracy of one part per trillion. And this isn't a "trivial" value, such as exactly zero or exactly one, but the exact number 1.00115965215, in perfect agreement with the measurement.

In terms of large distances and objects in the universe, Newton's theory of gravitation has explained the movement of the planets in the solar system since the 17th century, allowing us to predict their positions and also those of artificial satellites with great precision. Observations have allowed us to discover and understand more distant stars, galaxies, and galaxy clusters, too. Then, again at the start of the 20th century (in 1911), Einstein developed General Relativity as a theory of gravitation. This notion of space and time is completely compatible with the Special Relativity already used for objects (and particles) in labs. At the scale of the Earth and even the solar system, the transition to General Relativity is only manifested through small corrections to Newton's laws. At the huge scale of the universe, it describes phenomena visible through today's telescopes, such as gravitational "mirages." General Relativity has also made some extraordinary predictions, such as the existence of black holes, all of which have been confirmed by observations.

By juxtaposing these two big theoretical models — the Standard Model and General Relativity — we can describe the universe from the very smallest known elements to the largest observable distances. You may ask whether this combined theory is "right," i.e., if it has allowed us to make predictions, for example, and whether all these predictions have been confirmed by lab experiments or astronomical observations.

Unfortunately, the answer to that question is a bit complex: on the one hand, this big theoretical framework has met with great success; on the other, it has encountered some major difficulties.

Tallying the successes, the paired Standard Model + General Relativity comprises a remarkably productive theory. It describes the universe as flowing from the "Big Bang" at a given date in the past (estimated at 13.8 billion years ago), and not as a steady state. A large number of quantitative predictions are a direct result of this description, the most spectacular of which is the abundance in the universe of the simplest chemical elements (hydrogen, helium, lithium, etc.) — which observations completely confirm.

The difficulties themselves are of two kinds (to simplify things): on the one hand, a theoretical problem; on the other, a problem with certain astronomical observations. In terms of the first, the original two cornerstones — the Standard Model and General Relativity — seem fundamentally incompatible. In practice, they can be juxtaposed and the theories of each can only be applied in its original field: the Standard Model's Quantum Theory for small objects (particles, atoms); General Relativity for large ones (galaxies, galaxy clusters, whole universes). But look out if you try to extrapolate from one into the other's realm: applying General Relativity to particles leads to far-fetched predictions, or even no predictions at all. Likewise, asking quantum questions about black holes leads to difficult paradoxes.

The second problem is experimental: increasingly precise astronomical observations have pointed out major deviations from theoretical predictions.

Strictly speaking, a theory should be considered false as soon as a single one of its predictions has been contradicted by an experiment or, in the case of astronomy, by an observation. So why haven't we rejected the Standard Model–General Relativity altogether? The first reason is that we don't have anything else, or in any case, anything else that would succeed as well and with fewer problems. The second is that we invented some straightforward additions that can bridge the gap between prediction and observation: we dreamed up an unknown kind of matter, "dark matter," and an even less known kind of energy, "dark energy."

So we can choose between two attitudes. The first regards the theory as correct, saying that it has allowed us to identify types of matter and energy never observed on Earth in the lab. After all, that's exactly how Le Verrier discovered Neptune — by interpreting a deviation between calculations (in this case, Newton's laws) and the observed positions of the other planets.

The second claims that the price is too high to pay and would rather challenge the theory than accept the addition of these "exotic" components that have never been directly observed despite years of experimental efforts.

At any rate, the specter of Quantum Theory–General Relativity incompatibility lurks in the background. Are the two problems, theoretical and experimental, connected? Would a consistent theory also answer the question of dark energy and dark matter? Conversely, can more accurate measurements put us on the path to a better theory?

The New Conditions: Artificial Intelligence and "Big Data"

A third type of inquiry about our understanding of the world has recently arisen without invitation. It's obvious to everyone: the development of digital technology is changing our lives. A computer can perform tasks in our stead, ones that we had thought complicated, such as recognizing a face or driving a car. The general public is beginning to know about "algorithms," sets of rules that computer systems can apply in real time to classify, compare, identify, and evaluate. New ways of programming computers have appeared, which we refer to as artificial intelligence (AI), to use the general term.

The crucial step was introducing algorithms that can learn from examples, much as a human child does. Usually, a computer executes a series of operations (an algorithm) that human designers set in advance. To be sure, the algorithm can have "branches," executing this or that action if this or the other condition is satisfied or not. But this is largely mechanical: it could be done by a very large electromechanical calculator or even a completely mechanical machine made of cogs, ratchets, and springs. The algorithm's designers can test it in various situations and decide to modify it to improve it. But then it acts strictly as programmed. By contrast, an artificial intelligence algorithm can "learn" what it will need to do simply by studying a series of examples that say what we expect it to do for each example. No more need to specify the logic to use for each scenario. A computer programmed with an "artificial brain" can figure out on its own the logic it should use to "best" do what we ask of it for all the examples we provide it with. It will then apply the same logic to new situations, exactly as a human brain would.

There is a link between these methods and the increasing size of the data samples, i.e., "big data." Earlier, processing speed, data storage capacity, and speed and flexibility of data exchange, although sizable, were nonetheless limited enough that "conventional" algorithms could perform well. But they are less successful with very large heterogeneous datasets (images, texts, sound, physical or chemical measurements, GPS coordinates, DNA sequences, etc.), which AI algorithms handle beautifully. Indeed, this is where the human brain excels (or excelled): handling large amounts of very heterogeneous data efficiently and flexibly.

Articulating scientific laws, especially the laws of physics, is much like finding the logic of the observed world. If we gave an artificial brain *all* the results of *all* the measurements to date, would it be able to come up with the logic on its own?

Why would it have to show us its logic, in fact? Wouldn't it be enough that it could answer a precise question, such as, "If I make this measurement, what will the result be?" without revealing its internal mechanism?

In that case, what would be the difference between a "theory," as we normally understand it, and a black box such as this, if the two supplied the same predictions? And, to push this reasoning farther, could the black box do better than our current theories?

These questions may seem distant or abstract. But in my field, that of particle physics, we have been using AI algorithms on large data-sets for quite some time. We shall see how the change from traditional algorithms to AI ones is becoming increasingly profound and deeper, challenging the very idea of what a theory is.

Back to the Beginning: What Is a Physical Law?

It's quite possible that the crisis in physics will be solved tomorrow by the brilliant discovery of a new theory. It's also possible that the development of artificial intelligence will remain restricted to menial tasks, without its influencing scientific work. All then will continue as before — physicists will be proud of their latest theory, revere its author, and delight at repeating "Newton's" or "Einstein's" miracle. But in this case, the first crisis — the one between science and society — will be left out in the cold without being taken care of: increasingly abstract theories will be put to work in increasingly sophisticated technology, with the public's incomprehension and distrust still growing.

A contrario, we might consider that these three crises are different signs of one and the same problem — a crisis in how the physical world is depicted — and that their answers might be linked.

Therefore, this is a good opportunity to repeat the question — "What is a physical law?" — and go back to the very beginning. How do we create our representations of the world for ourselves? What is useful and why? What roles do our senses play? What do we call scientific intuition? After taking a step back from these questions, will we then see these "crises" as the end of physics or as a possibility for renewal?

My ambition for this book is not that it be a broad treatise on the development of human knowledge. The same questions are certainly being asked in other scientific fields as well. Nor do I aspire in this book to address all of modern science. I will simply try to delve into these questions by considering my field of research — elementary particle physics and cosmology.

Researchers find themselves confronted by these large questions daily. First off, of course, there is a crisis in physics, with all the theoretical and experimental efforts to try to surpass the Standard Model. Second, there is the major entry of artificial intelligence in extracting information from data collected from large experiments with particle accelerators (in this case, essentially the LHC at CERN). And lastly, there is a crisis between science and society, which I can see whenever I open a newspaper or visit a website, and which I do my best to ease by giving talks to the general public. Audiences, albeit self-selected amateurs, seem appreciative. Yet despite my efforts at popularizing in the truest sense of the term, so many attendees still tell me, "It's too complicated for me!"

Chapter 2

The Success and Limits of
Today's Physics

Obviously, it's impossible to describe modern physics in its entirety here, going from Maxwell's equations to superstring theory, and passing through Quantum Theory and Relativity. Many popularizing works do this task well. Rather, I would like to focus on a few concepts that have played a decisive role in shaping the representation that physicists have of the natural world. For example, with regard to small distances, the atomic hypothesis and 20th-century physics; for large distances, Newton's universal law of gravitation and Einstein's theory, including the Big Bang. Naturally, these are not the only examples of decisive progress in physics. Rather, it is important to realize that the success of certain ideas (we now call them "theories"), such as the above examples, has been so spectacular that they have left a strong mark on scientists: since then, physicists have frenetically sought to reproduce the methods and circumstances that led to the discovery of these ideas. At a time when physics is struggling with major obstacles, it's worth understanding how a model "that works" exerts a type of conditioning that is hard to escape.

Toward Small Distances

The Success of the Atomic Hypothesis

Human beings have always sought to understand the nature of matter around them, both out of curiosity and also so as to better be able to use it. As part of this search, two major opposing concepts can be distinguished. The first is an animistic view, or per modern terminology, "magic": each object, stone, tree, river, and so forth has its own identity; their relationships are completely context-based. As many "classes" and relationships between objects exist as there are objects themselves. The stories that people tell about these classes and their relationships can rely on metaphysical tales that involve higher beings, or if not, each object remains a tiny deity in itself. The second concept is reductionist: observed nature is built from a reduced number of stable elements, and the relationships between them are also regular and constant. The development of science and technology from the first tools up to the modern era has charted a course between these two concepts, often without worrying about any coherence between them. From the time of ancient Greece to the dawn of modern science, a certain type of reductionism has been at work, setting magical thought aside and trying to "explain" nature via simple models. But this has stayed essentially philosophical, with scientific and everyday practice dominated by empiricism. Until the 18th century, you could build bridges, purify iron, tan hides with soda, and manufacture cannons and arquebuses, all with the Aristotelian model of the four elements — earth, air, fire, and water — as the primary theoretical support. We cannot say that this model helped technological progress in any way.

Reductionism did not prevail until the 18th and 19th centuries, when it offered an understanding of nature that allowed us to create new practical applications well beyond empiricism. Philosophy yielded to theory; the experimental method replaced empiricism. For studying matter itself, the major turning point was the universal adoption of the atomic theory, which states that all matter consists of a collection of small atoms that exist in just a few types.

The atomic hypothesis was proclaimed in ancient Greece (traditionally, we attribute this to Democritus), which never ceases to be astonishing, as none of the parameters to support it were known at that time: there was no quantitative chemistry, nor instruments allowing one to "see" atoms.

Starting at the end of the 18th century, this hypothesis gradually became accepted, fully entering the mainstream by the end of the 19th century after a difficult journey disturbed by intense debates and controversies. Some scientists kept fighting until the beginning of the 20th century, particularly in France, even after Becquerel discovered radioactivity and Rutherford discovered the nucleus of the atom.

The impact and success of the atomic approach are so important for physics that it's worth briefly reviewing how it came about. Indeed, this more or less follows the development of the scientific model, which all physicists have in mind whether consciously or not.

The first approach that led to confirming the atomic hypothesis was via chemistry. Lavoisier first showed that most known substances could be formed from a limited number of basic substances, which became known as chemical elements, such as iron, sulfur, and oxygen. First and foremost, these substances were not connected to the idea of atoms, i.e., small elementary building blocks. They were solids, liquids, or gases, which known substances could be broken down into or rebuilt from. Such transformations between substances and components or between different substances are chemical reactions whose mechanism was unknown. But they could be quantitatively described by precisely tallying the amounts of the substances before and after the reaction.

With accurate measurements, Dalton and Gay-Lussac established that chemical reactions use fixed quantities of each substance and that the relationships between them are whole numbers. For example, to produce carbonic acid gas (today, we would call it carbon dioxide), you must combine 3 grams of carbon and 8 grams of oxygen, thus obtaining 11 grams of CO_2. If you add more carbon or more oxygen, it will be in excess and will not be combined. It's quite remarkable that the natural proportions in chemical reactions are whole numbers, here 3 to 8, and not, for example, 3.4 to 8.7.

Dalton interpreted these measurements by attributing to the simplest pure components (H, C, N, O, P, S, etc.) an integer "atomic weight," (1, 12, 14, 16, 31, 32, etc.), and showed that two by two, the combinations yield known substances, such as H_2O, CO_2, and NO. He immediately made an intuitive leap, invoking the atomic hypothesis and stating that matter is made up of atoms, "pure" substances have only one type of atom, and compounds have several types of atoms assembled together. This explains why combining a substance made up of A atoms with another made up of B atoms will always occur in fixed proportions.

Today we're accustomed to atoms and to classifying them by their mass and chemical properties via the so-called Mendeleev classification (the periodic table). But it's fascinating to realize that Dalton and a few other scientists (such as Berzelius) not only discovered that matter is made up of elementary particles, but they also discovered the periodic table of elements, which we now know as Mendeleev classification. Yet these concepts are not at all *a priori* identical! Why do atoms (at least the simplest ones) have a mass that is an integer multiple of the lightest atom, that is, a hydrogen atom? We now know that their nuclei all consist of a whole number of the same components — protons and neutrons. In today's terms, the reason Dalton's proportions are whole numbers or ratios of small whole numbers is not due to the chemistry of the atoms, but rather due to the structure of the nucleus of the atom. In fact, the mass of the atom is primarily the mass of the nucleus, as the mass of an electron is 2,000 times less than that of a proton. Every nucleus has a whole number of protons and neutrons, and the number of electrons in an atom equals the number of protons (an atom is electrically neutral). Moreover, the mass of a neutron is almost equal to that of a proton. The simplest element is hydrogen: its nucleus contains a single proton. Thus, it has become the unit of mass. The next element is helium. Its nucleus consists of two protons and two neutrons; therefore, the atom has two electrons. Since a neutron's mass is close to that of a proton, helium has a mass of four units, and so forth.

We could easily imagine a chemistry in which the basic, simple substances were "independent," instead of being built from the same basic components, protons and neutrons. Their atomic masses would have arbitrary, non-integer relationships, such as 2.718 or 3.1416. Under these conditions, it is highly unlikely that quantitative chemistry would ever have supported the atomic theory. Why invoke tiny building blocks if they exist in multiple varieties and if the rules for combining them are all different?

Likewise, if only the mass of the neutron had been significantly different from that of the proton, atomic mass would have been much harder to interpret! Imagine if the mass of a neutron were 0.785 times that of a proton (I chose the value $0.785 = \pi/4$ randomly out of ratios that are not "whole"). A helium atom, with its nucleus of 2 protons and 2 neutrons, would then have a mass $2 + 2 \times 0.785 = 3.6$ times the mass of a hydrogen atom. The masses of the atoms of H, He, Be, C, N, and O, which in the real world are very close to the whole numbers 1, 4, 9, 12, 14, 16,[1] would, in this precarious world, be 1, 3.6, 7.9, 10.7, 12.5, and 14.3, respectively. These numbers are a good deal harder to understand than the integers in our real world! Here is a case where nature has been kind enough to help our understanding.

The other route that led to the success of the atomic hypothesis came via thermodynamics, the branch of physics that describes the behavior of solids, liquids, and gases with respect to heat. Thermodynamics

---

[1]The fifth element in the periodic table, boron, isn't in this list because its apparent atomic mass (10.8) is not exactly a whole number. Boron nuclei naturally occur in two forms (isotopes): one containing five protons and five neutrons (a mass of ten); the other with five protons and six neutrons (a mass of eleven). We could still repeat the argument in the text by imagining all simple substances as a comparable mix of isotopes; their apparent masses would thus not be whole numbers and it would have been much harder to understand everything.

came into being when the steam engine was invented. In this branch of physics, we clarify the concepts of pressure and temperature, and we measure the characteristics of substances in connection with heat exchanges, such as their heat capacity, boiling, and melting points. An important mathematical framework allows us to easily manipulate the relationships between these quantities, thus making it possible to design complicated systems such as a steam engine or a refrigerator.

In the middle of the 19th century, Maxwell and Boltzmann realized that the laws governing the behavior of gases (first of all, the simple law of ideal gases) could be explained naturally if gases consisted of a large number of constantly agitated independent particles. The amount of agitation is measured by the temperature, and the pressure exerted by a gas on the inside of a vessel comes from the particles colliding into the walls.

This interpretation bolstered the atomic hypothesis that Dalton and chemists advanced: the modern atom emerged at the end of the 19th century. This was a triumph of reductionism, of concrete rather than philosophical reductionism. By observing nature on a human scale and making numerous measurements, we inferred its structure at a much smaller level. And as soon as instruments allowed us to conduct experiments, we discovered real elements that we had already understood would exist. This virtuous cycle is etched in the minds of all physicists and they keep reproducing it to this very day.

The success of the atomic hypothesis, however, was intimately connected to the fact that nature presented us with simple data, with elements whose atomic masses were simple, whole numbers and common gases whose behavior was very close to that of an ideal gas (we'll return to this). But who can say that all steps in physics research have to be so favorable?

From Becquerel to the Higgs Boson

We need not retell the amazing history from 1896, when Becquerel discovered radioactivity, to 1945, with the terrifying (and disgraceful) use of the atomic bomb on the Japanese people. This period saw the discovery of the atomic nucleus — protons and neutrons and all their properties, discoveries that bring us a remarkably consistent understanding of nature. Without going into any details, it's worth noting that nuclear physics explains the fact that the sun shines, in other words, that it produces energy and does so slowly enough to last for several billion years. In the 19th century, the sun's mechanism for producing energy was at the center of an interesting controversy: before the nucleus of the atom and nuclear reactions were discovered, the only known energy sources were chemical reactions, such as the combustion of coal. Given the power of the sun's rays and its mass, it was clear that if this power had come from chemical reactions — the only ones known at the time — a sun running on coal would have had to have completely burned up in several thousand years at most. We therefore arrived at a point where scientists — and first and foremost, physicists — contradicted the theory of evolution introduced by Darwin and developed by his successors! How could billions-of-years-old animal or vegetable fossils exist and how could species have evolved on Earth over hundreds of millions of years if the sun itself was only hundreds of thousands of years old? This is an example of science temporarily playing against itself.

The discovery of radioactivity and nuclear reactions came at just the right moment to explain the power radiated by the sun and its age. And this time, it was perfectly compatible with the age of the Earth estimated from geological and zoological evolution.

At the same time and in perfect symbiosis with these experimental discoveries, two major physics theories arose that disrupted our understanding of nature: Quantum Theory and the Theory of Relativity (the first part, known as "Special Relativity"). The first is traditionally associated with Planck, Bohr, Pauli, Einstein, and Schrödinger; the second is associated with Lorentz, Poincaré, and, of course, Einstein. I will return later on to these theories that clearly stated for the first time that nature — or at least the best description we have of it — does not behave the same way that our senses perceive every day. The very ideas of space, time, objects, and particles had to be rethought. With Quantum Theory, the very notion of reality moved from being an eternal philosophical question to a concrete, precise one. While there was no doubt at the time that Quantum Theory worked, nor is there any today, how Quantum Theory should be interpreted in terms of objective reality remains a topic of debate among scientists. Puzzling as these theories may seem, their applications quickly became essential in both the scientific world as well as the real one: for Relativity, nuclear energy, radiation therapy, and nuclear weapons; for Quantum Theory, electronics, molecular chemistry, and biology.

Continuing along the path of understanding nature at a microscopic level, we were able to reconcile Quantum Theory and the Theory of Relativity in two stages. Each of these began with a major problem, but each time, solving that led to an even more spectacular, important discovery.

Initially, simply describing an isolated electron within a theory that was both quantum and relativistic seemed impossible. In 1928, Dirac invented a mathematical formalism that permitted this description and wrote his famous "Dirac equation." By exploring this formalism as far as possible and particularly by extending his equation to concepts beyond the ones he had originally written it for, he discovered a solution in 1931 that described a particle with the same mass as an electron but with the opposite charge. This "positron," the first antimatter particle, would be discovered experimentally the following year. We currently use it every day for medical imaging (PET, positron emission tomography). Since then, matter and antimatter have been on equal footing in both experiments with particle accelerators and all our descriptions of the primordial universe.

We quickly realize that combining these two theories — Quantum Theory and Special Relativity — leads to completely unexpected consequences. Greatly simplified, Quantum Theory allows energy fluctuations if they don't last long. Relativity allows mass to be converted into energy and vice versa, and in particular, it permits a particle–antiparticle pair to be created from pure energy. As a result, for a very short period of time, a quantum energy fluctuation can produce a particle–antiparticle pair, as long as it recombines and disappears quickly enough. This is a simplified description, and the appearance–disappearance of the particle–antiparticle pair cannot be considered "real." Instead, we speak of a virtual process. But these virtual processes can, indeed, modify the properties of interacting particle systems. All these fluctuations and virtual processes in which even the numbers of particles are not conserved raise a concern that calculations could never give consistent, well-understood results. Nonetheless, we have managed to understand them. The first demonstration of the effects of combining these two theories was Lamb's measurement of a shift in the energy level in a hydrogen atom — a tiny effect, but one perfectly predicted by the theory.

This theory of matter and fundamental interactions, or rather, this theoretical framework, is called "Quantum Field Theory" and was developed starting in the 1930s. A sophisticated mathematical apparatus was established that allowed this theoretical framework to be used for calculating many quantities that could be measured experimentally. So even though the foundations of the theory are still uncertain — much like the foundations of Quantum Theory itself — physicists are amazed by its spectacular working success: all measurements made of atoms, electrons, and protons confirm the theory with extraordinary precision.

In the second half of the 20th century, technological developments allowed us to carry out research on the ultimate components of matter, using ever-larger means. We have, thus, moved away from studying spontaneous radioactivity or particles that arise in the cosmos naturally, to particle accelerators, whose sizes range from several meters during World War II to 27 kilometers for the LHC, the largest working accelerator in the world today. We discovered new particles, measuring their properties little by little, and understanding and interpreting their interactions in the same way that atoms were classified. Gradually, we classified elementary particles, completing their table and describing the structure of their interactions via some very specific adaptations of Quantum Field Theory, which everyone calls the "Standard Model." This model consists of twelve elementary components that interact with each other in two kinds of ways. Sometimes experiments have preceded theory in discovering the complete table of these components; sometimes theory has predicted the existence of new particles from existing data and they have been discovered later experimentally when a new accelerator allowed them to be created.

The most recent example is the discovery of the Higgs boson at CERN in 2012 via the ATLAS and CMS collaborations using the huge detectors at the LHC. The existence of this particle was predicted in 1964 (by Brout and Englert on the one hand and Higgs on the other) when the Standard Model was first invented and is, in a sense, its distinguishing feature. In retrospect, dreaming up the structure of the model and deducing the existence of the Higgs boson from the data available in 1964 remain an impressive intellectual tour de force. Starting in 1964, numerous measurements gradually confirmed the structure of the model, as did incremental completion of the table of components and their interactions: the Higgs boson moved from a mathematical curiosity to the focus of all research as each new accelerator was put into operation. Indeed, while the theory is based on the particle's existence, it says nothing about its mass, and hence the energy needed for an accelerator to create one. Before the LHC, detecting the "Higgs" was impossible. Thanks to its size, the LHC was the first accelerator with enough energy either to discover the "Higgs" and dramatically confirm the Standard Model or else refute its existence, thereby overturning the whole model. In 2008, after more than 25 years of construction and preparation, the LHC was commissioned. Despite a few initial problems, experiments since 2012 have proclaimed that the Higgs boson has been observed and its mass measured. Since then, we have made numerous measurements of its properties, all of which agree with the original theory and more recent developments of it.

Toward Big Distances

Newton and Universal Gravitation

The same reductionist approach has been used to understand humanity's surroundings at a long-distance scale. Rather than being a question of reductionism in terms of elementary components, this instead is a matter of confirming that celestial bodies can be identified and classified into a few types, as well as understanding their movements in the sky and also their relationships to each other. In doing so, we strip celestial bodies of their individual magic nature.

Among pioneers long before Newton, Giordano Bruno stated that the sun was a star and that stars were suns (which got him burned at the stake by the Inquisition). This assertion, even though it occurred in the middle of a complicated, dense, even fanciful speech, had all the power of the reductionist approach: formerly, stars could be thought of as pearls of nature that were all different, stuck onto a velvet vault — or entirely different charming images that weren't at all close to reality, nor able to be evaluated or contradicted. In just one sentence, the thousands of stars that could be seen in the night sky were both made commonplace and sublimated, becoming amenable to thought and reasoning. Obviously, the price to pay was that the sun was trivialized, moving from a special role for humans to being just one of the innumerable stars in the universe. In Bruno's time, the sun still revolved around the Earth and yet it had already lost this privilege.

Of course, Galileo, Kepler, and others measured and confirmed heliocentricism, but it was Newton and his discovery of the law of universal gravitation that gave this hypothesis a conceptual foundation and a practical way to apply it. Human beings have always observed the motion of the planets in the sky — those few "stars" that move about rapidly (from day to day and season to season) with a fixed background of stars between them. Their movements are regular enough to identify them completely but irregular enough that predicting them would be the subject of entire cults.

Again, confirming — as did Newton — that the motion of the planets and falling bodies is merely the same phenomenon is an eminently reductionist action. No longer did the moon and the planets need magic to move: their motion was just the same as that of falling apples.

Applied most simply, Newton's law of universal gravitation really allows us to predict the movement of the moon and the planets with great precision. Deviations from the law are quite small, caused by multiple interactions between the planets (still within a Newtonian framework) or by relativistic effects discovered much later. Here, too, the success of the theory is because nature presents itself to us as "straightforward." Our solar system has few planets, their speed is slow (therefore, the relativistic effects are minimal), their movements have the good taste to let themselves be explained by a very simple law, and as a result, a reductionist approach works spectacularly well. As with the atomic hypothesis, we can see that discovering the law was possible *because observed objects* — the moon and the planets — *confirm its simplest version with great precision*. I'm well aware that this sentence seems a lot like circular reasoning, but that does not in any way obviate the question of whether we would have discovered the law of gravitation under slightly more complicated circumstances. What if our system had two suns (something popular in science fiction novels)? Or if our planet orbited around a massive black hole, where we observed gravitation with what we today call "large relativistic corrections?"

Einstein, Current Cosmology, and the Big Bang

Once again, the goal of this book isn't to describe the development of cosmology since the start of the 20th century. Still, we should keep in mind a few pieces of information that will let us understand how our current representation of the universe was invented. Compared to the geocentric phase of magical or religious belief systems, heliocentricism merely describes what happens in the solar system, a stellar system like any other. But today, cosmology has acquired an entirely different goal: describing the whole universe. The universe has now become a topic of study, just like any other. Some theories posit that a great number of universes exist, with ours being just one example, a point I'll return to later on.

The crucial turning point was Einstein's discovery of the Theory of General Relativity, which remains the best example of a radical conceptual leap that has occurred in science due to a single person. Continuing to ponder space and time — a thought process that had led to Special Relativity — Einstein had the idea of including the effects of gravitation. Using nothing more than thought experiments, such as imagining an observer in an elevator or at the edge of a spinning merry-go-round, he grasped that the time a clock shows depends on the gravitational field it is in. For example, a clock hand doesn't move at the same speed on the surface of the Earth as it does on the surface of the moon, as their gravities are different.

He quickly discerned that *matter had to curve* space–time, i.e., the relationship between distance and duration is not only dependent on the observer's point of view (as was already the case in Special Relativity) but also on the presence of matter. The relationships are complicated. It would take him eight years and the help of mathematicians Grossman and Hilbert (and probably his wife, Mileva) to set up the mathematical apparatus that allowed him to manipulate these relationships correctly.

At Earth's scale, this new theory barely changed Newton's law of gravitation — changes that were, for the most part, hard to test experimentally at the time of Einstein's discovery. However, they've been measured thoroughly now, and they even cause a tiny correction in coordinates calculated by GPS systems. The first experimental proof came soon after the theory was invented by precisely measuring its effects at the scale of the solar system (an infinitesimal alteration in the motion of Mercury; an infinitesimal deflection of the light rays of stars by the sun's mass).

Next, naturally — if one can even say that, since we're dealing with progress by a genius — Einstein tried to apply his new theory to the whole universe, thus inventing modern cosmology: understanding the structure and development of the universe in its entirety.

General Relativity is an incredibly successful theory, as it predicts phenomena such as black holes and gravitational arcs (images of galaxies deformed by gravitation) that were completely unexpected but then proven by observation. Among all its implications, the most remarkable from the point of view of how we represent the world is undoubtedly the model of the Big Bang.

So as to avoid any misconceptions about the Big Bang, we can describe it as follows. Observations today show that the universe is constantly expanding. This does not mean that the universe is spreading "into a larger space." No, the universe is infinite. Its expansion is such that the distance between objects increases over time. A simple one-dimensional image can help with picturing this phenomenon: visualize a straight horizontal line in front of you, running left to right and with gradations marked on it. Imagine that this line is infinite in both directions, both to the left and to the right. Next let the line itself spread, simply by increasing the distance between the gradations, which should be easy for a line. Now take our universe with its three dimensions and do the same for all three dimensions. That is the universe expanding.[2]

During this expansion, the universe cools, much as all gases do as they expand.

Based on current measurements of the universe (distances, density, etc.), General Relativity allows us to calculate its development into the future and the past. So let's go back into the past by running the film of its development backward. As we rewind the course of time, the distances grow smaller and the temperature rises (as do all gases when compressed). We might imagine that the distances would become infinitely tiny and the temperature infinitely high only for an infinitely distant time in the past. But that's not what

---

[2] It's interesting to see that the public has no difficulty whatsoever with this one-dimensional expansion (a straight line), but finds it a good deal harder for three dimensions. We often get asked, "Where is the center point?" Or even, "What is the universe expanding into?"

calculations show us: the contraction isn't just proportional to the age in the past. As we go backward in time, the distances shrink more and more rapidly, becoming infinitely small (and the temperature infinitely high) not infinitely long ago, but a precise date in the past: 13.8 billion years ago. Thus, in the context of this theory, it is impossible to go farther back in time than then and the question of "before the Big Bang" doesn't make sense. An earlier date could only be meaningful by modifying the theory, for example, by inserting it into a larger framework so that it only becomes a specific aspect of the larger whole.

The simple idea that the universe hasn't been here for all eternity changes our concept of the world profoundly, directly contradicting our intuition — conditioned, as we are, by the apparent regularity of the seasons and the movements of the stars. But the Big Bang Theory does more than that: it lets us apply everything we know about physics at a microscopic scale, particularly nuclear physics, to the universe's first moments and to deduce completely measurable results. We can quickly cite the abundance of primordial chemical elements that were actually created in the first moments, or the existence and detailed properties of the "cosmic microwave background," radiation released by atoms about 380,000 years after the Big Bang.

The tremendous advances we have made in observing the sky from the Earth or satellites at different wavelengths have confirmed most of these predictions and allow us to measure the model's key parameters. While theory and observation agree well with each other, there are also a few discrepancies that have quickly become the principal problems in our physical model of the world, under the rubrics of "dark matter" and "dark energy."

Today's Big Questions

Physics today finds itself at the center of a huge paradox. Its theoretical and experimental advances have let us amass a body of knowledge of previously unattainable ambition and power. In practice, this base knowledge has allowed us to develop incredibly effective technology that has changed the lives of humankind — whether we find that good or not. Beyond such technologies, this knowledge offers human beings the completely novel possibility of thinking about the universe consistently and rationally without calling upon magical thought. Or rather, irrational thought has been pushed beyond the boundaries that rational thinking has itself clearly explained.

Basically, our ignorance is limited: in time, to the very first moments of the universe, fewer than 10^{-12} seconds after the Big Bang; in space, to not knowing the structure of matter at very tiny dimensions, smaller than 10^{-19} m. In principle, if our theory is correct, there are no restrictions to our understanding in terms of large distances. Within these limits, we should be able to interpret and understand all observed phenomena in the context of this theoretical framework. Any deviation is scrutinized and analyzed because it can cast doubt on the whole theory. We are therefore quite a long way from a theory such as Descartes' theory of vortices from 1644 (not so long ago), which attributed the movements of the planets to a never-observed ether that acted on the planets via "vortices," a theory with no experimental confirmation that also made no predictions that would let it be tested.

So it should come as no surprise that we have not completely achieved our Promethean ambition and that questions have appeared, big questions equal in size to our original ambition.

Dark Matter and Energy

The first big question crops up quite naturally from observations that do not agree with the theory. This should be the norm for how theories and our worldview evolve: if an observation is outside the framework, we must alter the framework or devise a new one. We'll see that the current situation is a bit more complex.

Dark Matter
In the 1930s, Zwicky observed an anomaly in the rotation of galaxies: the outermost stars rotate "too quickly," as if they are being pulled toward the center of the galaxy by a quantity of matter much greater than that of all the visible stars in the galaxy. The easiest interpretation is to imagine the presence of "dark" matter (dark in the sense of not shining like an ordinary star), detectable only from its gravitational influence on the visible stars. This solution duplicates a method that has already succeeded before, for example, in the famous discovery of Neptune, which was revealed via the perturbations it created in the orbits of the other planets.

Since then, numerous other observations have shown different aspects of this anomaly on a larger scale (galactic clusters) and in the evolution of the young universe after the Big Bang (measuring the cosmic microwave background). Since the model is supposed to be consistent, we have examined all the hypotheses about the nature of dark matter, but no known form of matter possesses all its characteristics. We have nevertheless proved that this cannot be an already known form of matter.

Dark Energy

At the end of the 1990s, observations of very distant bodies (luminous supernovas) showed that the expansion of the universe does not conform to the model arising from the Big Bang Theory. Apparently, the universe is expanding more quickly than the Big Bang predicts, and this difference is increasing with time. Nor is it a tiny correction: in our time (13.8 billion years after the Big Bang), the rate of expansion from the "standard" Big Bang is similar to the additional expansion from this mysterious cause. These measurements have since been repeated more precisely. Again, to interpret these observations, we must call upon an additional ingredient. As it cannot be matter (the presence of which would do the opposite, i.e., slow down the expansion), we've given it an even vaguer name, "dark energy." It turns out that there is a possibility in the Theory of General Relativity — a fixed parameter that Einstein called the "cosmological constant." After some doubts, this parameter was simply ignored from the 1920s until the end of the 20th century because the theory did a good enough job without it of describing an expanding universe to match our imprecise measurements from then. Only a few impertinent souls wondered why this parameter had to be exactly equal to zero. The discovery of *accelerating* expansion revived this well-known constant, which today we measure as a nonzero value in many different contexts.

But again, the microscopic theory — the Standard Model — offers no description of this kind of energy. Worse, attempts to calculate its density (the local amount) from equations that come from relativistic Quantum Field Theory end up with completely unrealistic values some 10^{40} times what we have observed.

In the case of both dark matter and dark energy, physicists are faced with a difficult choice: strictly speaking, one or the other of these observations would be enough to contradict the theoretical framework and lead to its being abandoned. But abandon it for what? Currently, there is no sensible alternative. And it isn't so easy to abandon a model that has achieved an extraordinary amount of success. Moreover, as far as cosmology is concerned, we have fairly simple descriptions available within the model — dark matter and dark energy — provided we ignore the gap with the microscopic theory. That's heartbreaking for us to do, since consistency between the micro and the macro, indeed the whole universe, is the foundation of modern physics' entire approach.

Problems with the Standard Model

Observations associated with dark matter and dark energy evoke the classic scientific cycle: a new experimental observation that contradicts the existing model leads to the discovery of a new ingredient or to the model being challenged. Yet even within the Standard Model of particle physics, there also are questions of theoretical consistency.

The simplest question is just, "Why?" In the Standard Model, there are twelve elementary particles, which can be grouped into three families of four particles each (two quarks, a "charged lepton," such as the electron, and a neutrino). Greatly simplifying things, the first family contains the particles of ordinary matter: protons, neutrons, electrons, and neutrinos. The other two families repeat the same properties (electrical charge, etc.), but with a different mass.

These components are (or are not) subject to two types of interactions, and each has a specific shape and mathematical structure. Clearly, the three families resemble a mini-Mendeleev periodic table, but we don't have the key to it. Why is the first family reproduced twice? Why these two types of interactions and what do they have in common?

Likewise, the model contains numerous arbitrary values (roughly 30) that are not predicted by the theory and can only be measured experimentally. These include the mass of elementary particles. From a theoretical viewpoint, the mass of these particles plays a similar role, but their experimental values are incredibly different: the ratio of the mass of the heaviest quark (the "top" quark) to that of the lightest quark (the "up" quark) is 100,000:1 and to the lightest neutrino, at least a trillion. None of this is theoretically impossible, but for a physicist, such differences are often due to an unknown underlying cause.

The other internal questions are more complicated. Without going into detail, they arise when we try to extrapolate the theory, via thought, far beyond what we currently can test experimentally today. We call one of these particular problems — a very technical one — the "hierarchy problem." Calculations show that the Higgs boson, which is so important in the Standard Model, is very sensitive to quantum effects, even much too sensitive. Every attempt to include the Standard Model in a theory that might aim to answer the above "why" questions, for example, would inevitably result in destabilizing the Higgs boson. This isn't a real inconsistency if you keep to the model, but it's an unpleasant situation.

Irreconcilable General Relativity and Quantum Theory: Superstrings

I mentioned earlier that the two main theoretical cornerstones of our understanding of nature, General Relativity and Quantum Theory, are incompatible. Let's try taking a closer look at where that incompatibility comes from, which has to do with the different ways that the two theories understand space–time and matter.

In Quantum Theory, space–time is mostly consistent with our usual concept of it. Space is measured by a regular, rectangular 3-axis grid of length/width/height. Time flows linearly at each point. Neither space nor time is influenced by the absence or presence of matter. To be sure, Quantum Theory allows matter to create fluctuations in the location, velocity, energy, etc., that characterize it, but it's easy enough to arrive at averages that behave "properly," namely, as our senses tell us. *On average*, the position of a particle is definite, as is its velocity. Considering the large number of particles, the *probability* of their doing this or that is quite well defined, and we can calculate it according to the theory. It even requires a good deal of work to engineer systems where purely quantum effects persist for large objects, long periods of time, or great numbers of particles. Even if they are full of potential fluctuations of energy and matter, these fluctuations average out to zero, space looks like space to us, and "large" objects appear well defined.

In General Relativity, the geometry of space–time is "flexible." The grid that measures location and date is malleable, and it becomes distorted when matter is present. We often use the image of a flat rubber sheet with an orthogonal grid on it to identify the points. Putting matter with mass upon it (a massive bowling ball) warps the sheet and causes the grid to curve. If we send a "test" ball onto the rubber, it will be attracted by the first mass due to the distortion in the sheet. We interpret its deflection as gravitational attraction.

In this vision, matter and energy don't fluctuate and movements are caused by the interaction between matter and geometry.

It's easy to see what makes combining these two descriptions challenging: for each point in space–time, Quantum Theory allows (and even insists on) a certain amount of fluctuation in matter or energy. Seen from the perspective of General Relativity, though, such fluctuations entail a local distortion of space–time. The coordinate grid itself becomes variable. The smooth rubber sheet becomes a kind of foam in which all shapes and sizes of hollows and bumps form and deform. It might have been possible that this maddening cycle of fluctuation → deformation → fluctuation could itself have been "averageable" and that when seen from afar, the foam would look enough like a sheet that clearly distinguishable coordinates and particles persisted. But that doesn't turn out to be the case: this type of cycle diverges. Each fluctuation entails an increasingly greater distortion and fluctuation. It becomes impossible to define an event in space and time, or a particle, or even empty space. We could almost forget that this "crisis" is only theoretical, a crisis in our representation of the world from two different, incompatible concepts. In nature, space remains empty and is kind enough to have an average of zero particles; gravity keeps exerting itself upon macroscopic objects without breaking its limits. The problem is only in the different *descriptions* the two theories suggest.

Of the many attempts we have made to reconcile these two approaches, the most famous is "string theory." String theory proposes that matter is not defined independently at each point in space, but that it should be seen as a collection of tiny vibrating strings. In a sense, the strings add correlations between possible fluctuations of matter at different points in space, thus limiting their tendency to increase wildly. Mathematically, the theory is quite complex; it would appear that it can only reach its goal of allowing a quantum description of gravity if space has ten dimensions instead of the customary three (length, width, height). Since we haven't observed any effect here on Earth that could show the presence of seven additional dimensions, we must assume they've been "compactified," i.e., twisted in on themselves into a tiny radius. To understand the idea of "compactification" in our three-dimensional space, you just need to picture a long tube. We can identify a point on the tube with two dimensions — the distance along the length of the tube and the position on the circle around that tube at that distance. This position around the tube is the compactified dimension. Seen from afar, the tube looks like a line and has only one dimension, length. Similarly, the additional dimensions from string theory would only be visible "very close up," at a very short distance, and undetectable with our current ability to probe.

After it was invented, string theory aroused a good deal of hope. We thought that its rigorous, precise mathematical framework would prove to be productive and even bring more results than what it had been invented to deal with. We dreamed that besides allowing a quantum version of gravitation, the theory would provide some answers to the large observational questions, such as those of dark matter and dark energy. Despite decades of effort, though, this is not the case today. It turns out that the model is not restrictive enough to be predictive: it isn't possible, for example, to decisively deduce the properties of known particles and interactions that are described by the Standard Model. Greatly simplifying things, there are an extremely large number of ways to "compactify" (twist) the additional dimensions, and in string theory we can create billions of billions of different types of universes. Strictly speaking, the theory isn't "false"; it is well defined mathematically. But it cannot make predictions or even answer questions such as "Why 12 elementary particles?"

The Paradox in Today's Physics

Since the beginning of modern science — say, since Newton — until the end of the 20th century, physics has advanced rapidly. Theoretical inventions and experimental discoveries have fueled each other, constantly bringing new ideas about nature and bestowing new powers for humanity to use. But for many decades now, basic physics has been faced with the abovementioned big questions. That does not mean the whole field has stagnated! At the same time, many promising branches of physics have continued to grow at a sustained pace, for example, quantum computing (I'll return to this later).

In particle physics, progress has been immense, too, even since the "big questions" appeared. But paradoxically, with every new real measurement and even every new thought experiment, the questions loom larger. We still have two theories, each incredibly productive and verifiable in its own realm, and, simultaneously, the same list of unresolved issues.

It's entirely possible that someone may come up with a totally new theoretical framework à la "Columbus' egg," replicating the legend of Einstein, with a new theory that suddenly unlocks all the mysteries. And it's entirely possible that a new measurement from an accelerator or a new observation of the heavens will send a clear signal about the nature of dark matter or dark energy, offering us Ariadne's thread to tug on in order to change the theoretical paradigm. In the meantime, we will have to arm ourselves with the courage to search and explore via observation and our minds.

Chapter 3

Understanding the World

Accepting or Understanding[1]

When we look at nature around us, we see all sorts of shapes. We can look at them and appreciate what they are without trying to classify or connect them to each other — in other words, without trying to explain them. For example, we might see the reflection of a mountain in a calm lake. We can simply see this reflection as one image among all the other parts of the landscape, the other mountains, the rocks in the foreground, and the plants, without asking any additional questions. But it's also natural to notice that the images of the mountain and its reflection have similar shapes while being symmetrical opposites of each other, and then to wonder,

---

[1]The example of a mountain reflected by a lake comes from the book *All of Physics [Almost] in 15 Equations*.

"why?" Why are these two images symmetrical? That's the start of the scientific method.

Finding an "explanation" for this reflection isn't easy. The standard answer today rests on the law of the reflection of light, and is expressed in the following way:

- The image of the mountain that we perceive through each of our eyes comes from all the light rays that leave the mountain and reach our eye in a straight line.
- The image of the reflection of the mountain, which we perceive through each of our eyes, comes from all the light rays that leave the mountain, which are then reflected on the surface of the lake before reaching our eyes.
- When a light ray reaches the surface of the lake at a certain angle of incidence, it leaves again (is reflected) at a symmetrical, equal angle.

If you don't have scientific training or if your experience of science is even a bit distant, could you express this interpretation of a reflection as clearly?

This example helps us realize that a physical law, here the law of reflection (the third point above), is only a small part of understanding, the "technical" part. What's most important is actually what's contained in the first two points, which involve several concepts that are not as simple or natural as one might think. First of all, it's important to understand that "seeing," isn't an active phenomenon but a passive one. We "see" because light rays coming from a source enter our eye. The very intuitive notion of a "look" is misleading. We do not "take" a look at a mountain or a person, no matter how dear the person looked at may be! At the most, we turn our detectors (our eyes) in its direction, thus permitting a tiny part of the light rays it emits to enter our eyes and produce a visual sensation. The commonplace sentence in a novel, "she felt his gaze on her back," is absurd, from a rational point of view.

In associating the direct image with the reflected image as coming from a common source, then, approximations arise. The surface of the lake isn't perfectly smooth, which creates distortions. The light rays that come directly to us from the mountain vs. after being reflected by the surface are not the same, and they may undergo different curvatures, absorption, etc., along their paths. Up to what point do we accept these approximations and still identify the reflection as that of a mountain? Of course, habit plays a large role: having seen thousands of reflections in all sorts of conditions, we identify a new one more readily. Further, having seen a reflection under ideal circumstances — such as in a mirror or a very smooth lake — helps us recognize the phenomenon of a reflection even under not-so-calm conditions, such as a lake with its surface ruffled by the wind. Once again, in order for me to identify a law of physics, nature first needs to show it to me in a simple, ideal form, one that is easy to codify.

This simple example offers us two lessons. First, that the scientific method is natural but by no means obligatory. Looking at a landscape, we can "accept" what we see without trying to explain it. We can be sensitive to it, find it beautiful or not, use it to know where we are, or use it to remember where to find food if we already have seen where it is. Or we can try to explain it. Second, that trying to "explain" is harder than one might think. Doing so involves a number of processes: classifying, simplifying, and approximating.

Explaining means finding approximate relationships between simplified objects that have been classified arbitrarily!

Animism, Monotheism, and Science

I am neither a theologian nor an ethnologist, but I am interested in cultures that use a different approach from that of science. The absolute opposite of science is the animist view of the world, where each object, every being, is considered "in and of itself," and inhabited by the same ontological power. There is no need to classify objects or beings, nor to simplify the relationships between them. As a result, human knowledge consists of recording what one observes in nature and then imposing upon it a metaphysical understanding that has no need of being verified by facts. This approach can be seen in the origin stories in animist societies. They often seem very complicated and lack logic (to us). But that's a bias in our culture: since animist societies do not simplify or classify their observations, the stories are quite close to nature, with numerous animals and plants, reproducing nature's complexity and its seeming arbitrariness. This knowledge goes hand in hand with an excellent understanding of nature and a large capacity for memorization. On the other hand, it is not conducive to developing science in the Western sense, with its rapid progress and ability to influence the world. In animist thinking, you don't interpolate: you don't put forward hypotheses about facts that are similar to those that have been observed. Likewise, you don't extrapolate; you don't put forward hypotheses about imaginary facts that would be a continuation of those that have been observed. So it is neither possible nor desirable to innovate, to change the world. There is no value judgement in this statement.

Is humanity happier by leaving nature behind and trying to control it using the scientific method, or by remaining a perfectly integrated component that works with the other ones, respecting them? This remains an open question.

In the West, the development of scientific thought goes hand in hand with "rationalization" as a metaphysical model. We move from the complete animism of early societies, to Egyptian, then Greek, and then Roman gods, with increasingly less specialized functions and increasingly less "magic" at each stage. We finally reach the generalization of monotheism, in which a single god is responsible for everything, along with the Bible's extremely boiled-down cosmogony in which the world was created in only six days. Again, each day consists of a "slice" that is independent of nature, with no more of a story than "God created this; God created that." I've always been struck by the simplicity of this genesis compared to the cosmogonies of other cultures — African, Native American, or Australasian.

Did this religious evolution favor the development of Western science? Or did the development of scientific thinking progressively eliminate gods from rocks and trees, and then from the sun, sea, and fields? What are the connections between these different paths? Discussing this isn't actually the goal of this book, but it's a question to keep in mind, specifically: is the theory of multiverses a new form of animism?

Scientific Consistency

A crucial aspect of the scientific method is consistency. If we use a model to describe nature or a small part of nature, then the model must account for *all* the observations in that domain. *One single* observation that contradicts the model is enough to dismiss it entirely. The way to advance is to come up with a model that agrees with all the already-made observations, find new situations that the model can make a prediction for, carry out an experiment or make an observation, and then compare the results to what the model predicts.

As long as this process involves generalities about the scientific method, this is obvious. Taken one by one, it turns out to be quite demanding and even harder than one might have thought. We all have our own knowledge from our studies, career paths, and encounters. Yet is this knowledge entirely coherent? Have we systematically examined all the relationships between its individual elements? Are there no contradictions between any of its constituent images or numbers? The breadth of our personal knowledge isn't necessarily what's important, but rather its consistency — if we can look at it dispassionately — and our ability to find these inconsistencies, examining new data, observations, or concepts critically.

Worrying about consistency has disadvantages. First and foremost, it's exhausting! If we tended to apply this requirement to every aspect of life, it would quickly become a cross to bear, both for us... and for everyone else. More seriously, this requirement tends to limit creativity. Do you have an idea for a new experiment to test a new hypothesis or measure an interesting number? Too often what happens is that you think, "No," this experiment isn't possible for this or that reason: too much background noise, inadequate sensors, exorbitant cost, etc. It's easy to be too rigorously critical and crush an idea or concept that's new compared to previous ideas or concepts. Therefore, it takes a genius to break free from the existing framework, overcome this critical attitude, and give creativity free rein, all the while maintaining overall consistency. It's very much a question of "letting go mindfully," to use an expression that's in today.

Alternate Facts

Opposing all kinds of unverifiable beliefs is not new for science — with religious beliefs first and foremost. But at the end of the 19th century and then during the 20th, it seemed as if the scientific approach was gaining the upper hand almost everywhere, that beliefs were retreating or at least that they were more clearly limited to an identifiable, well-restricted area. This was incorrect, however, as the advent of social networks has caused an explosion in the number and nature of "alternative facts," all kinds of unverifiable assertions based only on stories and acquaintances. Actually, the trend had already begun to reverse before social networks were started, with the rise of the creationist theories in the U.S., which began at the end of the 1990s. This reversal was undoubtedly connected to "too much science": the need to surpass the scientific approach, re-enchant knowledge — "enchant" in the sense of "bewitch" — and reestablish its magical aspect. Such resurgences remained relatively restricted in their reach in terms of the population that paid attention to them: for example, some religious communities (creationism), or relatively homogeneous groups (homeopathy). The boundary between known fact and belief was still well demarcated with rare (although famous) exceptions, such as the rumor that man had never walked on the moon. An image or sound was still hard to tamper with in practice and few controversies emerged, except for strong political reasons. But in those cases, we were warned to some extent.

Today, knowledge based on the scientific method is losing its singular position and becoming no more than one source of information among many.

With digital technology, manipulating an image or a sound is within everyone's reach, as is distributing it across the web, too. It only takes a few hours to get a large number of people to agree, a number comparable to a political party or a small church. Usually, such manipulations have a direct political impact. We also find many cases, however, where the political impact seems indirect or at least unclear. For example, there is a major stream of thought that maintains the Earth is flat (flat-Earthers).

The most disturbing aspect of these assertions is not their absurdity, but rather what appears to be a clear contradiction in logic. How can you insist that the Earth is flat when you use a satellite TV antenna every day? At a deeper level, there is a major sociological split here. Clearly, satellite TV works. But flat-Earth adherents completely leave understanding how this works "to others." To my mind, the problem is not that a fraction of the world's population believes and passes on these obviously false assertions. All knowledge has its limits and we are unintentionally disseminating a number of ideas that are either erroneous or that will turn out to be erroneous after scientific progress has been made. The problem is rather that this population is happy to use the results of science without coming to grips with it, entirely leaving the responsibility for that knowledge to someone else. In one breath, flat-Earth groups accept and defend an absurd belief, and they denounce the plot of a supposed ruling class or nations that would have us believe that the Earth is round. With the next, they completely subject themselves to telecommunications network technology. In so doing, they also subject themselves to the mindsets and markets found on those networks. And we cannot expect network operators to do anything to set the record straight. A call, a connection, or a click made by a flat-Earth group is still a call, a connection, or a click… I would very much like to see a disclosure on the screen of a satellite TV channel that states, "This is a satellite communications network based on the fact that the Earth is round and on Newton's and Maxwell's physics. To accept, click here."

Chapter 4

What Is a Law of Physics?

Modeling and Parameterizing: The Ideal Gas Law

To understand nature around us more effectively than by describing each individual phenomenon — and especially to let us use nature for our own benefit — we describe it with the help of the laws of physics. Generally, we start with observations or measurements that we have made of nature directly, as in astrophysics, or the outcomes of an experiment, that is, something that we have brought about ourselves in a lab. We identify patterns within the observations and measurements, much as we noticed the similarity between the mountain and its reflection. For example, for two quantities X and Y that varied during the course of an experiment, the easiest pattern to spot is a proportional one. This is what physicists call a linear relationship between X and Y, because on a graph that designates each experimental position by a point with the coordinates X and Y, the points are in a straight line. We are deeply influenced by this idea of proportionality, which feels so natural to us: biking twice as far takes me twice as long, three times as many potatoes weigh three times heavier, ten times as many potatoes cost ten times more, etc. Ah, but here the idea of nonlinearity arises: ten times as many potatoes often cost a bit less than ten times more.

A good many important physics laws are stated in the form of a simple proportional relationship:

- Law of Reflection, for a reflection in water or a mirror: the angle of reflection is exactly the opposite of the angle of incidence (a coefficient proportion of −1).
- Newton's Second Law, $F = m \times a$: Force F exerts acceleration a on an object. Acceleration is proportional to force and the coefficient of proportionality is called the mass of the object, m.
- Hooke's Law, $F = k \times X$: Force F exerts extension X on a spring. Extension is proportional to the force. The coefficient of proportionality is represented by k.
- Ohm's Law, $V = R \times I$: Voltage across a simple conductor is proportional to the current that travels through it. The coefficient is the resistance, R.
- For gases (gases called "ideal"): Charles's law states that the volume V is proportional to the temperature T (at a constant pressure). Gay-Lussac's law states that the pressure P is proportional to T (for a constant volume).

Of course, we sometimes find patterns between variables that are not proportional. For example, for gases, the Boyle-Mariotte Law states that pressure is *inversely* proportional to volume, that is, $P \sim 1/V$: the same gas compressed in *half* the volume has *twice* the pressure.

Combining the above three laws about gases yields the law of ideal gases: $P \times V = a \times T$.

I think it's important to emphasize here again the aspects of "modeling" and "parameterizing" in stating a physics law. Given a set of measurements, we defined some variables that seemed natural to us and we stated a mathematical relationship that linked them. These two aspects are inseparable from each other. Some variables are truly "natural" and others are less so; still others are only apparent when connected to a simple law. The ideal gas law offers an excellent example for examining the concepts of "natural" variables — or not — and modeling.

The Ideal Gas Law and Temperature

Before taking a closer look at the ideal gas law, let's first cover its modern interpretation, which was discovered in the middle of the 19th century. All bodies consist of atoms or molecules (groups of atoms). A gas is the form of a body in which its molecules (in the broadest sense, since in some bodies the molecule is a single atom) are relatively independent and move about in space in all directions. The pressure of the gas on the walls of a container comes from the collisions of these molecules with the walls. Temperature is the measurement of the molecules' kinetic energy (the energy contained in their movement). If a gas is somewhat diluted, the pressure is low (few impacts against the walls), there are few collisions between molecules, and we readily see that the ideal gas law is verified. In a smaller volume, there are more collisions with the walls and the pressure increases; a higher temperature means the molecules are moving around more, causing more collisions with the walls, thus more pressure. At a high pressure (we've put a lot of gas into the same container), the gas molecules occupy a large fraction of the space and collide frequently, which proportionately lowers the impacts with the walls, and the gas no longer follows the ideal gas law.

In order to be able to state this law, we must first have at hand the precise notions of pressure P, volume V, and temperature T. To be sure, these seem very simple, intuitive ideas, but are they accurate definitions? If we think about it, only volume is truly "natural." The idea of volume predates attempts to describe the physics of a gas by quite a long time; the concept has been used for centuries to assess the quantities of liquids and solids in order to describe them or for trade. Pressure is already a more difficult quantity: we can vaguely get the idea of it if we press on a soft surface of clay or wood with a smaller or larger cutting tool that leaves a smaller or larger mark. But it isn't so easy to clearly state that pressure is force per unit area. Force is fairly easily defined from its similarity to weight, but to go from there to defining the pressure of a gas in a container correctly and measuring it already assumes an entire intellectual and experimental construct. Understanding that we live in a gas — air — at a specific pressure, which varies with altitude, for example, is far more complicated still.

Temperature T, which operates in the law, is the most abstract of the three quantities. It refers to absolute temperature, in degrees Kelvin (K), where 0° is 273.15 K in our customary Celsius. Nowadays, our definition of temperature is based on our understanding in terms of statistical physics: temperature measures the degree of motion of the atoms or molecules in a body; 0 K corresponds to perfect stillness and a unit of temperature corresponds to a specific increase in the molecules' kinetic energy (their speed). But before modern physics, we could only establish a hierarchy of temperatures, i.e., more hot, less hot, with our sense of touch. Even when the law of ideal gases had been established, how to define a consistent temperature scale was not at all obvious. As a result, several scales were proposed, the most famous being Fahrenheit and Celsius. We now know how to convert from one scale to the other and can convert them into the absolute scale, K.

These original definitions of temperature highlight an interesting point, one that's tricky to explain. Let's focus on the Celsius scale for a moment. It is conveniently defined by its 0 degree point — the temperature of "melting ice" (a mixture of ice and water in equilibrium) — and its 100 degree point, the temperature of "boiling water" (a mixture of water and water vapor at atmospheric pressure). Fine, but once we've defined these two points, how do we define a degree? How do we divide the distance between ice and boiling water into 100 degrees? In order to set up a scale that allows any temperature to be measured, even between these two points, more than just the two endpoints are necessary. To better understand this subtlety, let's imagine I define a specific scale with 100 divisions between melting ice and boiling water. Nothing prevents me from modifying this scale by choosing larger intervals for the bottom of the scale (for example, by combining the original divisions two by two below 66°) and taking smaller ones at the top of the scale (for example, by splitting each division original above 66° in two): I would still have 100 divisions, but a temperature of 50° in the first scale would be 25° in the second one!

How, therefore, did Charles in 1787 and Gay-Lussac in 1802 conceive of their "laws" when they did not have the idea of absolute temperature available, but only the Fahrenheit or Celsius scales? Charles's law, $V = a \times T$, and Gay-Lussac's law, $P = b \times T$, do not work if you use one or the other of these scales directly. But they *almost* work. The relationship of volume and pressure as a function of the temperature in Celsius is, indeed, linear, but with a constant off-set: all you have to do is add 273.15 degrees to the temperature in Celsius (or 459.67 degrees to the temperature in Fahrenheit) to prove Charles's and Gay-Lussac's laws.

There are at least three "miracles" in this story:

- Real gases obey the law of ideal gases under ordinary conditions — around atmospheric pressure and room temperature.
- By carefully measuring the behavior of these gases between 0 and 100 degrees Celsius and then supposing that Charles's and Gay-Lussac's laws *should* be satisfied, we can *figure out* that an absolute zero exists, at –273.15 degrees Celsius.
- Dividing the distance between 0 and 100° Celsius works perfectly; to obtain the absolute temperature that will prove the equation $PV = a\,T$, you just need to add the figure 273.15 to the Celsius temperature. *A priori*, there is nothing to suggest this. Why wouldn't there be a slightly more complicated relationship, such as the following?

$$T_{absolute} = T_0 + a \times T_{Celsius} + b \times T_{Celsius}^2$$

In order to understand where this last miracle comes from, we need to go back to the experiments that led to the definition of temperature. Dividing the distance between 0 and 100 °C into Celsius degrees was accomplished in the beginning with a mercury thermometer, one identical to the mercury thermometers still in use today, at least until electronic thermometers were developed. A small quantity of mercury is enclosed in a tube and expands when heated from 0° to 100°C. The surface of the mercury travels along the tube, and graduations at regular intervals (in mm) create the temperature scale. Many bodies expand linearly with the absolute temperature. In other words, even before we determined the concept of temperature, we observed that if we heat mercury, another metal, air, or another gas, they all expand proportionally. And if, when we heat them, the volume of one increases by 10%, the volume of each of the others also increases by 10%. Thus, it is enough to use one of them as a reference in order create a standard thermometer and define the temperature scale.

But this similarity between substances as different as mercury and air is not at all as "natural" as you might think! It turns out that some bodies do not behave this way. For example, zirconium tungstate (ZrW_2O_8), *contracts* constantly when the temperature increases from 0.3 K to 1,050 K! Or, for instance, pure silicon contracts when moving from 30 K to 120 K and then expands at higher temperatures.

Let's imagine that all common bodies behaved differently with respect to temperature — some contracted, others expanded, and the rest had different tendencies at different ranges of temperature, even changing many times. Why not? Under these conditions, it would be very hard to establish a common scale for temperature, that is to say, to translate into a unique, universal quantity what our hand can immediately detect: more hot, less hot. To define a "good" variable, we must be able to perceive some kind of pattern — a pattern that is "broad" enough for the idea to make sense. What good would a thermometer be if it used a scale that was not ideal for any gas? That I couldn't use to predict the expansion of a single solid?

Or, if the relationship between volume and temperature in my thermometer changed directions at 39 degrees Celsius (102.2 degrees Fahrenheit), the indication, "37" (98.6 degrees Fahrenheit), could mean I was fine or on the brink of death.

Likewise, if our atmosphere were extremely dense, there would no longer be an ideal gas. Would we have been able to discover the law of ideal gases under these conditions?

Patterns in Nature: Scales

Essentially, we invent models that cover certain patterns that exist in nature. The big question is figuring out whether or not these patterns are essential and cover the bulk of natural phenomena. If so, nature would primarily be organized into large, regular patterns, with the rest being more chaotic and outside our ability to describe. On the other hand, if not, nature would be filled with irregularities: we could only find occasional little patterns here and there that we could model.

Modern physics leans toward the first possibility. It does seem that nature is largely organized into major levels of structures that are relatively independent from each other. Hence it is very important to understand that matter is made up of atoms. Luckily, the details of position, energy state, or spin are not crucial for describing the state of a piece of matter. Likewise, an atom contains a nucleus. Of course! But the atom's condition every instant does not depend on the exact configuration of its nucleus, and so on and so forth. Thus, there are separate *scales* — nuclei, atoms, ordinary matter, solar systems, galaxies, star clusters, universes — between which exchanges are usually fairly weak. Using a physicist's jargon, we would say that they are *weakly coupled*. At each scale, the one below looks as if it were averaged, with perhaps some signs of fluctuation, but nothing more. If the coupling between levels were strong, the slightest fluctuation in a nucleus (and there are many at the quantum level) would spread to the atom, to matter, etc. Not only would it be impossible to describe this state, but it would undoubtedly be impossible for matter as we know it to exist, that is, with varied but stable structures. Everyone knows about the "butterfly effect" on the Earth's atmosphere: the atmosphere is a system with strong couplings between scales. A tiny fluctuation at a small scale rapidly spreads to a higher scale, and so forth. This is the famous example whereby a butterfly's wings flapping in Brazil sets off a hurricane in Texas. Fortunately, the whole atmosphere is a fairly closed system, and its interactions with the rest of the Earth are weak. Thus, this "chaotic" situation does not spread. Imagine if electrons, nuclei, atoms, matters, stars, and galaxies were strongly coupled. Then everything would be chaotic. The different scales would blend together all the time, and it would be an easy bet that none of the structures could appear, as bonds would be made and unmade all the time at every level.

In any case, this is a modest but interesting demonstration of the anthropic principle: we exist because nature is organized into structures of size (and energy) that are separate and relatively independent.

Inside one of these "strata," the laws of physics are relatively simple, proportionality rules, and we can hope to understand them. Our task as physicists, therefore, is first to discover these structures and study how they are organized. Once this foundation is laid, we can explore the points of contact between these different levels, and the nature and extent of their communication with each other. So physics will then seem more complex and less intuitive, in that our intuition is fed by the appearance of a stable, linear nature where effects are proportional to causes.

What's a *Good* Law of Physics?

We can see what the ingredients of a good physics law are. The quantities it describes are well defined and relatively universal. The law is simple, or at least as simple as possible (unfortunately, the laws in modern physics are anything but simple!).

We could write thousands of laws that would only be narrowly valid, that would only apply to specific objects and under particular conditions. A law such as that would be preceded by a long list of conditions under which it would be valid.

For object X, placed this way, in such and such a place and time, then quantity A, which is a very complicated function of the size, weight, and temperature of object X, and quantity B, which is another very complicated function of its chemical composition and hardness, verify law A (a very complicated function of B), provided that A remains between 0.99 and 1.01.

Clearly such a law would not be a *good* law of physics. That's what we call "Occam's razor," named after the 14th century theologian and philosopher: the simpler a law is, the better it is and the more its discovery will permit progress. Why is that, after all? Why are we always happy to adopt a simpler law than the preceding?

The classic example is heliocentricism. Before Copernicus, the Earth was believed to be fixed and the other planets moved around in the sky. Simple measurements of their positions show that they do not simply revolve around the Earth. A planet that moves in the sky for hours or weeks, according to an apparently simple trajectory, can stop moving, go the opposite direction (retrograde) for several days or weeks, and then get going forward again! So it quickly becomes evident that the planet does not circle the Earth at a uniform speed. Rather than questioning the Earth's immobility, we built a famous (famous for its inanity) theory of epicycles: there is indeed a point that describes a circle around the Earth, but it isn't the planet itself. The planet revolves around a "virtual" point. Obviously, since we've added an extra element that can be adjusted (in physics jargon, we say that we have added an additional "degree of freedom"), our description of the planet's apparent motion will improve. But little by little, as the measurements grew increasingly precise, this model no longer was adequate. Therefore, we added a circle around the previous point, and so on, the result being the "theory of epicycles," the textbook case of a "bad" theory, where successive refinements only served to mask its fundamental error, the hypothesis that the Earth doesn't move. All modern physicists have this lamentable scenario in mind: what if the complexity of modern physics were just a new theory of epicycles?

We know the following: in Newton's theory, two very simple laws — the fundamental principle of dynamics ($F = m\,a$) and the law of universal gravitation ($F = -G\,m\,m'/d^2$) — can, on their own, explain the motion of the planets with impressive precision. The first, the fundamental principle of dynamics, can be applied much more broadly, as it states how *all* bodies react to *all* types of force.

How do we go from the "geocentric + epicycles" theory to that of "heliocentric + Newton's laws?" Historically, this is the whole story of Tycho Brahe, Copernicus, Galileo, Kepler, and finally Newton. But here, let's just be aware of the following fact: when Copernicus proposed heliocentricism (that the Earth and the planets revolve around the sun), his hypothesis did not immediately solve the problem. This is because the simplest heliocentric hypothesis is that the Earth and the planets go in *circles*, with the sun at the center, and this is not the case. We would later discover that they move in an ellipse, with the sun at one focus. Therefore, we had two concurrent descriptions for some time: one — geocentricism + epicycles — experimentally precise but completely "false" conceptually, and the other — heliocentricism + circular orbits — much "truer" but less in agreement with observations! It would take dogged persistence with measurements and a good amount of intuition for Kepler to discover that the heliocentric model "worked" perfectly, but with elliptical orbits, and for a specific relationship between the position of a planet in its orbit and its speed along that orbit, the famous "law of areas."

At each stage in developing the laws of physics, we encounter the same dilemma: keep the old laws by refining them, thereby rendering them increasingly less "attractive," or risk turning everything upside down at the cost of a period of uncertainty, even complete failure!

All the same, one might wonder why Occam's razor works so well. After all, we manipulate numerous complicated equations and we also use innumerable specialized, practical parameterizations, each specific to its own field, which make no claim to being attractive. Will it always be true that the "simplest" laws are the most productive?

The primary reason for the "efficacy" of laws that are the simplest to state probably is that the human mind is limited. Our short-term memory is restricted: on average, we can remember roughly seven one- or two-digit numbers. Our mental processes are even more limited: we believe we are very intelligent because we can learn about Quantum Theory or General Relativity. Yet think about the game, for example, where we have to respond to a list of questions by moving the answer forward by one, in other words, where it is necessary to bring forward the answer to the previous question. Without practice, this simple game is hard for us, whereas the minimal intellectual gymnastics it requires is completely elementary for an algorithm.

To create a theoretical construct, a representation of the world, we need simple structures, ones that are easy to memorize and use.

To be sure, we've compensated for our limited brains with a whole set of tools to help us think and remember. First of all, there is the invention of writing, which allows us to retain many more values and concepts than short-term memory. But writing is a double-edged sword: on the one hand, it mitigates our memory deficits, but on the other hand, it weakens our memory because we are no longer forced to use it. The same goes for the invention of the internet, for the same reason.

As luck would have it, nature also needs simple structures, as we saw above. Therefore, there is a good chance of our simple minds finding a good description of nature via simple structures.

That is what a *good* physics law does, ever since people first tried to conceptualize the world that surrounds us.

Hence, one question arises at this stage. What if there were a tool that allowed us to surpass this limitation of the human mind, that let us manipulate large quantities of data and understand overall relationships without having to organize that information to make sense of it? In short, where is artificial intelligence taking us in this context? We'll come back to this as well.

Domain of Validity, Extrapolation, Boundaries

Of course, as soon as a *good* law of physics is available, it becomes interesting to explore its boundaries. First, in the interests of being honest, accurate, and pragmatic, it's important to understand the law's "domain of validity." While physicists can delude themselves that their law is universally valid, engineers must know the point beyond which it no longer works. A cable can stretch in proportion to the weight that hangs from it (Hooke's law), but past a certain weight, the cable stretches much too quickly and breaks.

Second, it's also important because the boundary — the limit of its domain of validity — will often prove to be interesting. Let's take the law of ideal gases as an example again. Maxwell and Boltzmann discovered that the atomic hypothesis (that gases comprise independent molecules) explains the law for dilute gases, where the molecules almost never collide. But what happens at higher pressures? If the molecules have a non-zero size, they will collide more frequently. Collisions between molecules will cause there to be fewer collisions with the walls of the container, i.e., pressure, and the law of ideal gases will no longer apply. In 1865, in a tour de force of both physics intuition and mathematical ability, Loschmidt, using Maxwell's work as a starting point, succeeded in connecting the viscosity of a gas (a small deviation from the law) to the size of the molecules, deducing from it for the first time the approximate size of an air molecule to be a few billionths of a meter (a nanometer). This was such a small value that it was completely beyond what instruments of the time could directly observe.

Exploring the boundaries of a law often happens experimentally with extreme temperatures, pressures, currents, voltages, etc. Such exploration can also be done purely intellectually: we just ask ourselves the question, "What if…?" What happens to Newton's law of attraction between two bodies at a very short distance, when force (which increases as distance decreases) gets quite large? It even became a scientific method in its own right after the invention and development of quantum mechanics at the start of the 20th century. This new method of describing nature goes against intuition so greatly (we'll come back to this), that in order to better explore it its founding fathers vied to design thought experiments (*Gedankenexperimente*). In Quantum Theory, concepts and objects are vague and misleading. Only the results of calculations count. Thus, this kind of experiment involves describing the initial conditions as precisely as possible, next performing the calculations using thorough, inflexible quantum methods, and then trying to interpret the results. The more the results of the calculation turn out to be counterintuitive, the more interesting the *Gedankenexperiment*! One of the best-known examples (to physicists) is the Einstein–Podolsky–Rosen paradox, which only existed in the form of a *Gedankenexperiment* for a long time. The paradox has to do with the behavior of a certain number of photons which, at the end of the day (after the calculations), would be in several different places or would arrive without having been said to leave, would communicate with each other more quickly than the speed of light (a limit for photons), etc. Anyway, without going into the details here, at the end of the calculations, those devilish photons, even if only imaginary, behave in a way that puzzles physicists. This single *Gedankenexperiment* has caused more ink to flow than countless real measurements. It wasn't until 1982 that researchers, with Alain Aspect at the forefront, managed to reproduce the conditions in the laboratory that Einstein, Podolsky, and Rosen had imagined, transforming the *Gedankenexperiment* into a real one. You will not be surprised to hear that the results were perfectly compatible with Quantum Theory: the theory is right, but we don't always "understand" it!

A law of physics can also turn out to be unexpectedly productive. Invented to account for a series of measurements or prior knowledge, it then proves to encompass much more than what it was initially designed to explain. I'll take two examples that physicists know well: Maxwell's equations and Dirac's equation. Maxwell's equations describe the behavior in time and space of electrical charges and currents, and of electrical and magnetic fields. Maxwell wrote about them in 1865, synthesizing several previous laws — those of Ampère, Faraday, Gauss, and Kirchhoff. Maxwell's equations bring together electricity and magnetism in a powerful way, allowing countless systems to be calculated and developed — electromagnets, electrical motors, dynamos, transformers, the telephone, etc. But if you really know how to read them, his equations also describe the generation and propagation of all electromagnetic waves. Radio waves, which were completely unknown when Maxwell wrote his equations, are one example of this. Thus, far from only being a parametrization of the known, this law of physics opened the door to an immense, rich, unknown realm. Even better, toward the end of the 19th century, we would see (particularly, Poincaré, and Lorentz) that these equations entail a curious entanglement of space and time, an alteration of distance and duration when traveling near the speed of light. This would be the starting point for the Theory of Relativity, which Einstein discovered in 1905.

The second example is Dirac's equation, which he wrote in 1928. At that time, Quantum Theory (also discovered in 1905) had already made considerable progress. In fact, we had just detected one of the properties of the electron experimentally — spin — which Quantum Theory had predicted and which there really is nothing analogous to in classical electromagnetism. Relativity, on the one hand, and Quantum Theory, on the other, seemed to work perfectly well, but there still was no common theory that could connect the two essential aspects of elementary particles and forces. This, then, was the problem that Dirac addressed. With extraordinary creativity, he wrote a simple, elegant equation that contains these two aspects for the electron. Remarkably, the equation described the recently discovered spin. With that, Dirac's equation would already be an essential foundation for particle physics. But it gets even better. Dirac realized that the equation had two solutions. The first depicted an electron with all its already known properties. The second solution seemed to depict a strange electron that would go back in time, or else whose energy would be negative. Far from ignoring this extra solution, Dirac observed and analyzed it. It would take him three years to find another interpretation of it, one totally revolutionary at the time: the second solution described a particle with the same mass and spin as the electron, but with the opposite charge. If it collided with an electron, this "anti-electron" could even annihilate it, resulting in pure energy (actually, two or three photons). The following year, Blackett, observing the interactions of cosmic rays in the atmosphere, found an electron with the opposite charge among the particles. Thus, Dirac's "positron" was, indeed, real. We would understand later that all particles have their own antiparticles, a particle with the same mass but the opposite electrical charge (as well as other charges, as the case may be).[1] Today's physicists study and manipulate particles and antiparticles together. Positrons are even used today for medical examinations — for PET imaging, positron emission tomography.

Thus, Dirac "read" the existence of antimatter in his equation, which could not have been surmised before his law. Dirac's equation would become the point of departure for the theory that underpins all modern particle physics, "Quantum Field Theory."

---

[1]Truth be told, some neutral particles, thus without any charge, can be their own antiparticle.

Chapter 5

Accurate, False, Incomplete, and Natural Laws

Induction and Falsification: Popper

O bviously, we would like our physics laws to be "true." But what exactly does "true" mean in this context? A mathematical theorem can be true, in the sense that it can be proven logically from well-defined axioms. (We'll set aside the caveats added to this phrase by modern mathematics, and, in particular, Gödel's notion of incompleteness.) For science — for physics — that's not the problem. The problem is just that a law of physics should take into account natural phenomena whose variety and extent are not defined as premises of a theorem. After observing 50 black crows, can I conclude that all crows are black? This question, which philosophers call the problem of "induction," has been the subject of long debate, which there's no need to reproduce here. A famous joke illustrates the problem in a few words:

An astrophysicist, a particle physicist, and a mathematician arrive at JFK airport in NYC for a conference and go to catch a taxi together. All the ones that pass by are the famous yellow cabs. Once he's observed the first two or three, the astrophysicist takes the plunge:

— In this city, all taxis are yellow.

The particle physicist waits a bit longer, and then after seeing about a dozen of them, goes one step farther:

— In this city, most taxis are yellow.

The mathematician waits even longer. Finally, he declares:

— In this city, there are at least 23 taxis with at least one yellow side.

Who is right? Is the most correct law the most useful one?

In 1953, Karl Popper expressed this problem in its simplest form: we can never "prove" a scientific law, a law that describes nature. Even if we have carefully identified its variables (the physical quantities involved), and carefully defined its domain of validity, a doubt will still remain. For example, experimental measurements made with today's precision may seem to confirm a law. But if measurements become more accurate tomorrow, they may, of course, reveal a deviation — minuscule, to be sure, but undeniable — with regard to the law as stated today.

On the other hand, according to Popper, we can refute a physics law, or falsify it. A single experimental measurement that contradicts a value predicted by a law is enough to falsify the whole law. Even if, overall, the law seems to describe the reality of nature well, with a single recognized contradiction from an experiment, no matter how isolated, peripheral, inessential, or even far from the principal meaning of the law it may be, the entire law is called into question.

Therefore, when physicists propose a new law, they should, in principle, verify that this law is compatible with *all* known experimental measurements. Taking this rule at face value, it would thus be useless to propose a law that only described one part of the data or that described all of it very approximately, but the details of which were contradicted experimentally.

All the same, taking this rule at face value might be a bit harsh and could lead to stifling creativity: a new idea doesn't immediately coalesce into a well-defined, well-ordered law that can describe all the measurements and observations at first blush. We saw that with the example of heliocentricism: fortunately, we did not abandon the idea just because circular orbits did not describe the movements of the planets accurately!

When facing the large problems we have alluded to, we can say that anything is allowed in modern physics — all new, interesting ideas, even if they don't directly solve these problems. It's enough if the idea is original and that it perhaps proposes, with many conditions, a slender thread that might eventually lead to a hint of a solution to one of these problems — even if that idea says nothing about other problems or even makes them worse. In fact, the more the problems seem fundamental and insurmountable, the more we're ready to pursue ideas that are far from predictive or even "testable" models. That's the criticism made by some people about "string theory," at least in its original version, where the scale of the size of the strings was so small that they would be far beyond our ability to measure or observe.

Between False, Incomplete, and Untestable Laws

So are there "truly false" theories, and which are they? I would say that there are, indeed, but that this isn't the case for most of the theories falsified by Popper's criteria: strictly applying it would be far too rigorous and would not allow any innovation. Let's put to one side short-lived theories based on blatant experimental errors, such as "Blondlot's N-rays," from around 1903, or all the theories that appeared after the (erroneous) measurement of faster-than-light neutrinos in 2011. For the rest, a theory proves to be false when a number of measurements at the very core of its focus contradict it. The theory then loses its raison d'être. What was the use of geocentricism, even loaded with numerous epicycles, once Newton's theory was established? What good was the continuous theory of matter once we discovered atoms, radioactivity, and electrons?

We quickly forget "false" laws. On the other hand, a great many important laws have turned out to be incomplete rather than false. This is the case with Newton's laws of mechanics. To be sure, the theories of relativity (Special Relativity for mechanics and electromagnetism; General Relativity for gravitation) have completely changed how we think about space, time, force, etc. The fundamental concepts in relativity are very different from those in Newton's laws. But for everything that isn't too far from human senses — speeds that aren't too fast; distances and times that are neither too short nor too long; moderate gravitation — Newton's laws remain excellent approximations. Indeed, it is easy to show that the laws of relativity approach Newton's laws under these conditions. Even so, the fundamental concepts of Newton's laws could be considered "false," as they presuppose instantaneous interactions at a distance, whereas relativity and experiments dictate that the transmission of an interaction cannot be faster than the speed of light. However, since Newton's laws are a good approximation over a wide range, we consider them incomplete rather than false.

Similarly, Fermi's theory (1934) of weak nuclear interactions constituted a big step in understanding elementary particles and their interactions, allowing us to comprehend and quantitatively calculate certain disintegrations in radioactive nuclei. Thus, the model was immediately useful for both basic research and also, for example, for designing the first nuclear reactors. However, we realized soon enough, at the end of the 1950s, that the model couldn't be the last word on this type of interaction: just calculating the probability of two high-energy particles interacting (much higher than what experiments of the time allowed) resulted in a probability above one, which makes no sense. We therefore had an elegant, effective new law of physics that we quickly knew was incomplete, which could only represent the low-energy limit for a better theory that would be valid across a much broader domain.

For that matter, most researchers feel that our current laws are imperfect and incomplete, but not really "false." Instead, they will be the low-energy limit (or for distances that aren't too tiny) for a future theory. Obviously, it's impossible to choose between these two hypotheses; we may be sinning out of vanity, and our current laws will look totally false to physicists in the future, with the new concepts opposing the old ones and rendering them obsolete. Or we may discover a new theory that will resolve all of today's problems, with our current laws kindly considered approximations.

Under these conditions, how do we sort through new ideas? How do we choose which will have any chance of emerging as a testable, falsifiable theory if we devote the energy needed to develop them, and drop those that will not lead anywhere? What we do know today is that comparing an idea to experimental data immediately will no longer suffice. Indeed, most physicists do not believe that a new Einstein or Newton will discover a new comprehensive theory where all deductions using already measured quantities will agree with the data (within measurement error) and will suggest new measurements that we can test it against using current or feasible instruments. Given the enormous number of measurements already made in the context of modern physics across extremely varied systems and a range of energy of greater than 20 orders of magnitude, discovering the magic theory, the universal equation, in a single step seems inconceivable.

Hence, theory is being challenged. How far do we go to develop a concept that never will be "testable' (such as string theory, for example)? Or even one that will only solve one very specific aspect of the problems? (For example, this is the case for models that only try to connect the three families of particles using simple symmetry). Should we keep a theoretical hypothesis that seems capable of bringing us answers but has not had the slightest experimental confirmation across decades?

This last is the case of "supersymmetry," an attractive theory (rather, theoretical framework) that might solve some of the internal problems in the Standard Model and provide a candidate particle for the dark matter in the universe. But it says nothing about dark energy and has even more free parameters than the Standard Model. Most importantly, in its most "natural" version, it predicts the existence of new particles, basically a partner for each ordinary particle. These "supersymmetric" particles would be endowed with even greater masses than that of their classic partners, which is why they have not been discovered yet. But for this partnership to work, the masses of the supersymmetric particles should not be too different from the masses of their Standard Model partner, which should make them within reach of our experiments with the large accelerators. Naturally, since physicists are enterprising optimists, they always find reasons to think that these potentially tremendous discoveries are "just around the corner." As a result, we have searched for these particles every time a new accelerator was commissioned without ever finding them. That doesn't mean that the theory is *false*. All we know is that it is incomplete and has not yet received specific experimental confirmation, i.e., of one of its explicit predictions.

Naturalness

To sort through the proposed theories, or to be able to explore current theories and try to identify their weak points more comprehensively and specifically than just stating their major problems, modern physicists are very interested in the concept of "naturalness." This involves making "Occam's razor" more precise, more quantitative. As already mentioned, a theory with a limited domain, and that has to have a number of parameters adjusted to account for reality, is not particularly attractive. It will be even less so if its entire structure critically depends on the exact value of these parameters. Supersymmetry, just mentioned earlier, is one such example. The theory predicts a set of new particles as partners to ordinary particles. To satisfy this new symmetry, the mass of each super-particle should be equal to that of its ordinary partner. As we have not observed a single one of these super-particles, we deduce that their mass will be greater than that of their partners, and thus out of reach of our accelerators for the moment. The symmetry we just invented has, thus, been "broken!" That isn't a problem in and of itself; we know of examples in which broken symmetry that is just "as it should be" is a good start for a theory. For example, that is the basis of the Brout–Englert–Higgs theory, which allows for the Higgs boson and the whole Standard Model. But in the case of supersymmetry, this "symmetry breaking" occurs at the expense of the theory's "naturalness." In the original symmetric theory, we had the same choice of masses and strength of interactions as in the standard theory. When symmetry is broken, the choice is limited: if we want broken supersymmetry to continue to do its job — which, let's not forget, was to solve certain problems in the Standard Model — we must now choose to adopt some very specific parameters. Gradually, as we push the limits for super-particle mass, we have to concede that the other parameters of the model will become adjusted more and more precisely. We move away from the initial idea of natural, aesthetically pleasing symmetry where super-particles and particles have the same mass. Currently, assuming super-particles are just around the corner, and just beyond the limits of current research at the LHC, the parameters would need to be adjusted by a few percent. Is that too artificial or is it still "natural?"

In any case, can we objectively assess "naturalness?" Why should an adjustment of one in a thousand seem far-fetched to us? We accept ratios much larger without flinching: for example, the ratio of the mass of the heaviest quark to that of the lightest one is around

100,000, and yet they play exactly the same role in the Standard Model. And the the ratio of the mass of the heaviest quark to that of the electron is four times greater still. The Standard Model doesn't predict the values of the masses of elementary particles. Rather, a single mechanism generates their mass: the Higgs field interacts identically with all particles but at different strengths (couplings). Is it "natural" that this same source would produce effects with such diverse intensities, adjusted only by the greatly differing values of the free parameters?

When we consider a collection of free parameters that play similar roles in a theory or a model, we are used to saying that they are of "the same order of magnitude" (less than a factor of ten between them), "somewhat different" (a factor of 1,000), or else "very different" (a factor of 1,000,000). "Very different" values for two parameters that play the same type of role intrigue us. We would like to find the reason for this imbalance. Is the smaller parameter canceled out by a fundamental principle (conservation, symmetry)? Is the larger one reinforced by an effect (resonance, exponentiation...)?

But does nature have the same sense of orders of magnitude as we do?

Indeed, our sensitivity to differences in values undoubtedly comes from the simple fact that we look at orders of magnitude in powers of 10 — because we have always counted on our ten fingers. For a centipede, the idea would not be the same. More seriously, some people shrug off these questions by hiding behind the anthropic principle, saying that these values are what they are in our universe, they would be different in another, and it's pointless to look for unknown physics laws in connection with them.

Chapter 6

Modern Physics and Intuition

Representing Nature

Doing physics means building mental representations of the world around us. To be sure, we can describe these representations, give them written support (equations, books, websites), convey them (classes), and use them to create objects or perform actions. But all of these are constructs that come from human brains and are manipulated and used by our minds. So the connection between the natural world and our brains is key, whether it's at the level of "hardware" (our sensory organs, neurons, etc.) or "software" (our social interactions, culture, etc.). Before the "modern" era (the 20th century), the representations adopted by physics were relatively easy to imagine and visualize.

Science — and first and foremost, physics — began by describing and conceptualizing phenomena that our senses could observe directly. Next it addressed phenomena that our senses did not have direct access to. But we still used images with analogues in direct observation. This direct relationship to reality supports theoretical descriptions. Even if you are embedded in Aristotelian culture and can only conceive of a universe centered on the Earth, it's "easy" to change your mindset and picture the Earth and the planets revolving around the sun. We can imagine changing our viewpoint from one representation to another, as both are equally "intuitive." Thus, this step was much more cultural than intellectual.

Physics then moved toward the abstract. For example, Newton's and Fresnel's optics, which describe light as waves, already require effort to go beyond the most immediate intuition. We intuitively see light as a ray, a straight line in space, and now we must replace that image with a wave, a wave that we cannot perceive directly. At most, we can observe its effects via the interference phenomena that led to our adopting this model. Luckily there are many wave phenomena that our senses absolutely can perceive, such as acoustic waves that we can visualize and produce via vibrating strings in musical instruments or waves on the surface of water. Moving from the image of a "ray" to that of a "wave" is a bit distressing — it already creates some tension between perception and conceptualization — but for scientists and engineers, this tension is "acceptable." Note, nonetheless, that this tension is already enough to separate the world of those in the know from the rest of the population, even in the most developed countries. Despite centuries of education, who today besides practicing scientists really imagines light as a wave and from that correctly interprets the colors we see in an oil slick?

This movement toward abstraction more or less continued in 19th century "classical" physics, even though it was already very effective.

The major turning point came at the start of the 20th century with relativity and most of all, Quantum Theory.

The Question of "Reality" (Quickly Dropped)

Obviously, in order to ask the question of representing what's real, we would first like to make sure that reality exists. Does reality exist? The question is old, much like philosophy, but unlike science. For "classical" scientists, the question does not arise. They investigate, discover, and understand the nature of a real world. With the invention of Quantum Theory, the observer gains a more important role, as they can never be isolated from the nature they observe. This challenges the intuitive model of nature as something that is what it is and that we see as objective observers, and rekindles discussions among scientists about the existence of reality: we can think that everything is merely the product of our imagination, ourselves included — a sort a *Matrix* worldview. Sure, but in which fishbowl?

We quickly arrive at questions that are more metaphysical than physical, and just as pointless. I do not think the crises in physics are directly related to this kind of questioning. After all, even if "reality" is nothing more than self-generated thought, that only intensifies the representational crisis: why shouldn't we formulate coherent, effective representations of a reality that our mind would itself create? So let's drop this question, and like all good scientists, admit that "reality" exists, or at least, let's act "as if" it did. But be careful! While the idea of reality can be very broad, this does not mean we are returning to classical physics.

The Question of Instruments

At my talks for the general public about particle physics or the Higgs boson, the question of our instruments keeps coming up, as in, "You can't see the things you're telling us about; how do you know that they really happen the way you say?" I admit that our accelerators and detectors are complicated and that we need increasingly large computer programs to reconstruct the collisions of protons and trace the information the detectors provide us back to the particles, their type, charge, and energy. But how is that fundamentally different from the instruments used in all other branches of physics? I tell my audience that we took the first step in this direction centuries ago with the first astronomical telescope and the first microscope; since then we have interposed an instrument between our sensory organs and nature.

Here again, I do not think the "crisis" is due to the complicated technology in our instruments. A 19th century lab would have looked complicated to a non-expert, with test tubes, mirrors, lenses, electrometers, and galvanometers. But such complexity has never seemed to create a stumbling block for explanations — quite the reverse. A modern quantum optics measurement made with lasers, radio frequencies, and individual photon detection that gives us access to this mysterious branch of physics is no more artificial than measuring the electrical force created by rubbing a rod with cat fur.

All representations spark "imagination," in the sense of creating an interior image, one that today we would call "virtual." Images, therefore, play a special role in constructing physics theories. When the measurement itself consists of providing an image, it acts directly as a representation of reality. Hence optical instruments hold a privileged position, as if they rendered reality more faithfully and more credibly than other instruments, much as images have long held a favored status for conveying information. The notion that "a picture is worth a thousand words" is decidedly difficult for human beings — with human eyes — to renounce. Yet the most beautiful images that modern telescopes and microscopes produce are computer-generated, using filters and detectors that have nothing to do with human vision. Still, I envy the ease with which my astrophysicist friends can talk about their science to the public while relying upon these images, whereas I have the greatest difficulty explaining microscopic physics, especially trying to avoid analogies that are easy — and false. Giving in to such ease, all scientists (including particle physicists!) try to popularize their fields via images, even though their "real" results are not expressed that way but rather via tables of figures or graphs that are much more visually "neutral."

Oddly, it is possible that the widespread usage of computer-generated images may wind up correcting our cognitive bias for visual proof. Where everyday information is concerned, it is quite easy today to create images that look completely real. A "deep fake" video can thus present a politician saying or doing anything with no difference between the fake video and one shot from a real scene. When we have become used to the fact that all images and audio recordings can be completely synthetic, we will have taken a large step toward ridding ourselves of preconceptions about scientific instruments. Paradoxically enough, the public will then be able to derive *more* confidence in the scientific method from this: since what I see can be modified, manipulated, and even totally made up any which way, let's instead trust statements that come from instruments and methods that adopt the scientific policies of transparency and reproducibility.

The Turn of the 20th Century: Relativity and Quantum Theory

Since we have already brought this up a number of times, it's time to see how these two theories — the foundations of modern physics — defy our intuition.

Einstein published his theory of Special Relativity in 1905. As mentioned earlier, it was unknowingly contained in Maxwell's electromagnetism, and curious, logical minds, such as those of Lorentz and Poincaré, had already traveled along the same path. But they had run into the wall of "common sense." To use the simplest example, in relativity, the speed of light does not depend on the reference frame where the observation is made from, and it cannot be exceeded. A simple thought experiment immediately shows the paradox that "common sense" leads to. You're in a car that's moving at velocity v_1. You throw a ball forward out the window at velocity v_2 for you. Your friend who has stayed on the road measures the speed of the ball compared to the ground. Obviously, she will find that velocity $v = v_1 + v_2$. If the automobile experiment doesn't work for you, just realize that running to gain momentum when throwing a javelin, for example, allows the speed of the javelin to be faster relative to the ground than when throwing the javelin while standing still.

Let's return to the car. When it's stationary with the headlights on, the light moves forward at the speed of light, which is traditionally notated as, "c" (for celerity). If the car is driving with the headlights on, does that change how quickly the light travels with respect to the ground? Common sense answers, "$v_1 + c$." But relativity answers, "still c," and experiments confirm this.

It's amazing that the fact that the velocity composition law is not simple addition constitutes such a difficult paradox for those who have not studied relativity. This is the source of numerous audience questions and of a number of manuscripts I've received from sincere, passionate amateurs who want to prove that "Einstein was wrong." However, all we need to do is accept that the law of composition of velocity is not $v = v_1 + v_2$ but instead is

$$v = \frac{(v_1 + v_2)}{\left(1 + \dfrac{v_1 \times v_2}{c^2}\right)}$$

for the paradox to disappear.[1] The formula is not simply additive; but it is not very complex either. It shows that although the speed of light has a set value (in mathematics, we'd say that it has a "finite" value), it is also a kind of infinity for velocity: if we add a specific number to infinity, we always get infinity. Why do we have so much trouble accepting this type of "composition?" We have come back to the tyranny of addition and "linear" composition that I spoke of above.

This odd composition for velocity is just the superficial sign of a more serious challenge to the relationship between time and space. A more significant sign, for instance, is that a clock on a spaceship appears to run less quickly when viewed by a stationary observer on the ground than it does for passengers on the spaceship. In addition,

---

[1]In particular, if v_1 or v_2 is equal to c, then v is also equal to c; for v_1 and v_2 that are far less than c, this is reduced to simple addition, $v = v_1 + v_2$.

the length of the ship when observed from the ground is shorter than when measured on board the ship. These effects are larger, as the speed of the vessel with respect to the ground is a larger fraction of the velocity of light. The speed of our spaceship is a small fraction of "c," but extremely precise atomic clocks can easily measure the effect on time. Such "relativistic" effects are at work for high-energy particles manipulated in particle accelerators and experiments. For example, many particles are unstable, and we can measure their average time to decay when they're "at rest," i.e., when they're immobile in our lab. The same type of particle at almost the speed of light takes much longer to decay than when it is at rest. These slightly odd laws of movement (kinematics) have been confirmed and exploited.[2]

Lastly, we cannot bring up relativity without mentioning the equation $\mathcal{E} = mc^2$. I do not want to get into detail here about what it means. But in the context of this book, it is interesting to note that if we interpret it correctly, this equation really poses no problems in how it represents the world. On the contrary, it puts an end to debates that raged long before Einstein about the relationship between inertia and energy. The fame of this equation and the public's lack of understanding of it do not come from its strange or revolutionary nature but rather from its practical developments in nuclear physics — first and foremost, the so-called "atomic" bomb (actually a nuclear bomb). Other equations, other prior theories, and centuries of chemistry had allowed us to develop an

---

[2]Langevin made the strangeness of these laws famous in his twin paradox, but a detailed interpretation of the paradox and its solution turned out to be less obvious than Langevin thought.

arsenal that was just as deadly — attested more than enough in World War I — but that was part of global development. Dynamite was invented for construction as much as for shelling, and the chemistry of poison gas was only a tiny part of industrial chemistry at the start of the 20th century. The invention of nuclear weapons constitutes the first — and up to now the only — time that an advanced theory based on cutting-edge lab research enabled us to design a weapon of considerable military and political importance, even though no parallel development had reached civilians (other than radiography using X-rays and a few risky attempts at radiotherapy). This, therefore, has justified the fantasy of an absolute weapon that will render its first inventor omnipotent, and it has contributed to the public's alienation from physics even more.

In 1911, as Einstein continued to question space and time, he discovered the theory of General Relativity, which we already spoke about at the start of the book and which is the basis for all modern cosmology. As mentioned earlier, the theory proposed a groundbreaking vision about gravitation, and some of the effects it predicts are spectacular — black holes, gravitational mirages, gravitational waves, etc. But its effects around us are small and the transition between our distances and cosmological ones takes place "smoothly." Beyond the original difficulty with the relativity of 1905, which we call "limited," once "general" relativity arrived on the scene, nothing more really collided with our intuition. In other words, the new theory did not contradict our intuition about "continuous," "incremental" effects, in which a small increase in a cause produces a small increase (or decrease) in its effect. In studying a black hole, the weirdest object that the theory predicts and describes, we often thought that the frontier (the technical term is "horizon") displayed a "singularity," a place where the theory failed, as the relativistic effects become infinite. We then proved that was not the case, and that with a good mathematical description, everything once again went "smoothly." Only the center of the black hole is "singular," but no more so than Newton's gravitation if it is extrapolated to infinitely tiny distances.

In short, General Relativity is a theory that feels "acceptable" to our intuition, despite profoundly changing our concepts of time, space, gravitation, and even the age of the universe. Intuitive representations are easy to find — such as the deformed rubber sheet — and not all that distant from mathematical truth.

Things are quite different for Quantum Theory. Even though this group of concepts has been developed, studied, and used for practical purposes for over a century, it retains a mysterious, paradoxical nature. Niels Bohr, one of the fathers of the theory, said, "Anyone who is not shocked by Quantum Theory hasn't really understood it." And Richard Feynman (who won the Nobel Prize in Physics in 1965 and was a great architect of Quantum Field Theory as well as a professor well known for his teaching ability) went one farther, saying, "I think I can safely say that nobody understands quantum mechanics." You might imagine that these both are cases of false modesty and irony, but they are not. If you know the work of these two great scientists, you understand that they are instead carefully weighed statements.

As soon as you start studying Quantum Theory, peculiar behaviors leap to the eye. The very first exercise I had to do in my very first course consisted of calculating the probability that a particle placed at the bottom of a hole would spontaneously move to a neighboring hole. After very simple calculations, you find that this probability is not zero. It's tiny, of course (if the hole is deep), but it exists as a possibility, whereas this situation (a particle passing spontaneously from one hole to another) does not exist in classical mechanics. You can immediately see that Quantum Theory is not a slightly modified extrapolation of classical mechanics and, thus, from the intuition our senses provide us. It's "something else."

Along the way, objects lose their materiality. The concept that best describes particles then is that of "wave functions" rather than material points, i.e., a number of material points in space that show the possible presence of a particle. This "fuzzy" image of matter has caused a good deal of discussion, but I don't think this description presents any particular representational difficulty. We have been using electrical and magnetic fields as well as temperature fields and speeds of a fluid for a long time.

Ideas that are truly new and hard to conceive of are linked to how systems evolve. The first strange evolution occurs when we measure a particle or any system of particles. Unlike in classical mechanics, in Quantum Theory, measuring always affects the system. So it is tempting to include the instrument of measurement within the system being studied and to calculate their shared evolution. One thing leads to another, it seems necessary to include the observer, and the difficulties begin. What to do about the observer's free will? It appears as if the measurement depends on the observer's choice, even when they do not act upon the experiment and even when it seems materially impossible that they could have acted before the experiment had ended.

The paradoxes quickly multiply — and these are "hard" paradoxes that have resisted decades of study. Not that the theory is false or even imprecise. On the contrary, the formalism, if we accept it, always allows us to calculate the result of an experiment. The most famous is certainly the Einstein–Podolsky–Rosen paradox, which was invented in the 1930s and was only a thought experiment — but one whose logic seemed to necessitate major challenges to the idea of causality (cause coming before effect) or locality (no instantaneous influences at a distance) or both. Despite Bohr's opposition, Einstein tried in vain to explain the paradox by imagining the existence of "hidden variables" — unknown properties that would guide the behavior of particles. At that time, and continuing for years, we knew about this paradox, but we could still ignore it and believe it was a purely theoretical problem or perhaps a problem of the language we were using. But in 1964, Irish physicist John Bell made it "impossible to avoid." He created a very realistic thought experiment based on photon measurements, with decisive results. Only Quantum Theory was able to pass the test; all attempts to describe the experiment using classical mechanics — even when equipped with all the hidden variables one might want — could not. As we saw above, that experiment has since been carried out by Alain Aspect and then repeated with increasing refinement and precision. The result is always in favor of Quantum Theory.

Make no mistake, the positive results of this experiment "explain" nothing. They simply show that it agrees with the description of the system and the calculation of its evolution in a quantum framework and not a classical one. The major paradox is, therefore, that we know "how to do it" but we don't "understand it." At any rate, it all depends on what we mean by "understand." After all, how is a working description based on calculations any less of a good understanding than a discussion based on our intuition, which is, itself, essentially created from our sensory experiences? It is clear that our senses only allow us to perceive a tiny fraction of natural phenomena. To be convinced of this, just compare the range of wavelengths of light that our eyes can see (from 0.4 to 0.8 microns) to the range of all electromagnetic rays that we know how to use with ease today (from well under a nanometer to well over a kilometer). It is quite possible that our brain, which handles information from our senses, is also adapted to modeling nature at our own scale, but that it is not structured to be able to model nature at other scales of distance or time, which forces us to invent theories that seem complicated and paradoxical.

A Few Neurological/Psychological/Sociological Questions

I am not going to settle the above question of whether the human brain can "imagine" Quantum Theory here. But I would like to draw your attention to some strong limitations that our senses impose on how we think about and see the world. As a first example, we have already mentioned the theories that try to reconcile gravitation and Quantum Theory by postulating the existence of additional dimensions. Mathematically, it's easy to use more than the regular three (length, width, height). Instead of locating a point with three coordinates, we just need to add the desired number of dimensions. We find "vectors" in high school classes — lists of coordinates in n dimensions, adding them, increasing them by a fixed number, rotating them in n-dimensional space, etc. None of this poses a problem for a professional physicist. We can talk to a theoretician about her latest discovery constructed in ten-dimensional space, in which our world is a three-dimensional sub-space and in which material particles and forces gallivant around in ten dimensions.

But I always find it amusing that for this theoretician to illustrate her point, she goes to the blackboard (or often now a whiteboard), and what does she draw? A perspective view of one or two little corners of sheets of paper, with several lines connecting them. The blackboard (and its perspective) is supposed to show ten-dimensional space and the little corners show three-dimensional space such as ours. In short, even for a professional researcher with all her abilities of abstraction, the best way to "see" the problem is still to reduce it to our regular vision. Our world is in three dimensions and we only have two eyes to see it with, which gives us a (small) perspective. Mathematicians have found numerous properties of space and objects with more than three dimensions that have no equivalent here. Physicists frequently use these properties to construct their theories. Despite our knowing that these properties can neither be reduced nor projected onto our type of three-dimensional space, our limited habits and senses remain extraordinarily important.

Transmitting ideas also calls for simple, even simplistic images. For successful communication, language simplifies concepts and sometimes even uses purely sociological conventions that have nothing to do with the concept but allow us to quickly get our bearings. Here's a small, surprising example of this. Before we had digital and computer projections, we used "transparencies" for conference presentations. For those who have never experienced this rudimentary technology, this is an 8 1/2″ × 11″ sheet of transparent plastic that you write on by hand with color felt-tip pens. To illustrate a presentation, you put this sheet on an "overhead projector," which projects it onto a large screen for the audience. So we used to write and draw on the transparency by hand. (A story within the story: many of us spent the night before an important presentation correcting transparencies in our hotel rooms. To do this, we would erase the marks with toilet paper soaked in eau de toilette, as felt-tip ink was alcohol soluble. The next day, when we placed the corrected transparency on a hot overhead projector, it emitted a pleasantly perfumed fragrance). Thus, it often was necessary to represent particles on the transparency and we had to quickly choose between colors for each type of particle. (Since digitalization, the diagrams we draw on the computer do not offer the same spontaneity.) I realized then that a vague convention had become established for the colors used for each particle. Electrons were often blue or black, protons red or brown, and a great majority of muons[3] were drawn in green. However, no written or oral convention existed, and you could easily find counter-examples. Actually, these particles have no color, neither in the regular sense nor in any theory. So where did this convention come from? I think it came from our strong need to identify with an everyday object or material, even by experienced physicists who play with these particles every day.[4]

Thus, from our direct senses to our most sophisticated culture, our psychological environment shapes our way of thinking about the world. Our well-developed scientific theories exist at the outermost periphery of our network of connections and influences. And it

---

[3]The muon is a very well-known elementary particle that physicists use frequently. It's similar to an electron, but in the second family of particles.

[4]We could explain these color choices by an association between a particle's "intuitive size" and the wavelength of the color in the visible spectrum: the smallest (Compton wavelength), the electron, matched to a short visible wavelength, hence the color blue; the largest, the proton, corresponding in the visible spectrum to red; and the mid-sized muon matched to green. This interpretation is even more astonishing, since physicists would have made this association unconsciously without it ever being formulated.

requires quite an act of contortion for our intellects to access phenomena they absolutely were not trained to perceive via natural development. Only "forced" development, e.g., through education, allows this access. Under these conditions, it's not very surprising that from time to time, our intellects get stuck along the way. This, perhaps, is the simple reason why modern representations of nature in physics appear counterintuitive and for the apparent crisis running through them.

Chapter 7

Can We (Really) Change Paradigms?

Unexpected Discoveries

On occasion, an experimental discovery (or a series of them) requires us to question concepts, all the while offering us new paths. This was the magical situation at the turn of the 19th to 20th centuries. At the end of the 19th century, "classical" physics was at its zenith: mechanics, electrodynamics, and thermodynamics formed a powerful, coherent, consistent whole from a working perspective. Other sciences shared its success: chemistry and medicine were quantitative, rational, and also functional.

At the time, most scientists felt that physics was "complete," that there would never be a need for new concepts. Yet Michelson, who had just proven that the speed of light is independent of the reference frame it is measured in (which appeared in obvious contradiction with classical mechanics of his time), said that "the major principles have now been firmly established. All that physicists will have to do now is measure to the sixth decimal place." This quote is often and unfairly attributed to Lord Kelvin. Poor Kelvin serves as an easy scapegoat here because despite his remarkable contributions to science (absolute temperature, the Joule–Thomson effect, Ampere's definition), he also was quite unlucky in his predictions and assertions (that machines heavier than air could not fly, that the sun was 20 million years old, etc.). But this time, although he, like others, was convinced that physics was essentially complete and perfect, he mentioned that its success was obscured by two dark clouds: the fact that no one had observed "ether," the mysterious fluid that bathed our world and allowed light to spread, and the difficulties in interpreting the spectrum of atoms from the partition of energy according to Maxwell–Boltzmann, a theory that was, however, logical and had other successes.

Kelvin's speech about these "two dark clouds," which he delivered to the Royal Institution of Great Britain in 1900 and subsequently developed, is still well known. In point of fact, the first "cloud" would be explained a few years later by the Theory of Relativity and the second just as quickly by Quantum Theory.

But in 1900, few scientists worried openly about these problems. At most, the two questions were in the background of more cautious questions. Rather, it was the avalanche of experimental discoveries especially that led us to question these concepts completely: the discovery of X-rays, different types of radiation, photoelectric effect, etc.

Everything was miraculously in place for a scientific revolution. First of all, classical physics had provided a solid base: quantitative and quite predictive, it allowed theory to be compared with measurements, bringing attention to disagreements that required further examination. Next were the sophisticated mathematical developments: linear algebra, group theory, curved space, etc. Finally, there also were the newly mastered experimental methods: photography, electricity, chemistry, etc.

Without minimizing the worth of the men and women of science of the time, one can almost say that at that moment, it was enough to let oneself be carried from experiment to discovery and new experiment to unveil a reality that had been completely unknown up to then. All of today's elementary particle physicists dream of living in a period like that. But we have already carried out numerous experiments and made countless measurements, and the new ones we can envision are primarily large scale, taking time to implement and analyze. It is unlikely that we will get another rapid series of experiments undertaken over a short time that will lead to a conceptual revolution as easily as connecting the dots.

As a result, we examine the data from all our experiments to detect the slightest "Kelvin dark cloud" that might open a path to new understanding. Admittedly, we see those large clouds that we have talked about, looming far in the distance — dark matter, dark energy, the quantum/gravitation incompatibility — but we lack the smaller, nearer clouds that can be tested experimentally. And we eagerly await the day that a path will spontaneously appear from the results of one of our major experiments.

To Really Challenge Concepts

Part of the myth is that significant progress in our understanding will always be linked to challenging concepts at their core.

We understand the planets' motion because we abandoned geocentricism and put the sun in the center. We understand atoms and particles because we abandoned our regular notions of time, space, and material bodies. What must we abandon today? And what will replace it?

While we're on the subject, I think we might find it interesting to question the role of mathematics. Normally, we've congratulated ourselves that as physics has evolved, mathematics has always provided the necessary tools for the latest physics theory of the day when needed. Eugene Wigner was fascinated by these achievements and came up with the famous question of where the "unreasonable efficacy of mathematics in the natural sciences" comes from. In fact, examples from the past abound: conics for Newton's gravitation, differential analysis for classical physics, group theory for Quantum Theory, Riemann's curved spaces for relativity, etc. But are we really sure this will always be the case? Admittedly, mathematicians have been incredibly creative in exploring structures, relationships, and geometries, but who says the tools we need to get past the current crisis are available?

We have seen how our minds are shaped by our senses and intuition. Functional relationships, especially linear ones, completely predominate our methods of analyzing a problem: *a little more* cause leads to *a little more* effect. It's difficult to escape this kind of thinking all at once, because we experience it every day and because this kind of *differential* analysis has seen such success for the sciences in the past.

There are numerous developments geared toward escaping this way of thinking. Nonlinear phenomenon analysis is a major field in physics. It describes what happens when water freezes, for example: what are the conditions under which water begins to solidify? The phenomenon is very much nonlinear, since a slight lowering of the temperature makes liquid water change to a completely different state, solid ice. Water changing into ice is an example of a phase transition, a transition between two very states when a parameter, here the temperature, changes just a little bit. The transition is a collective effect: the atoms in a quantity of water, which were disordered, suddenly arrange themselves into the form of a crystal. Right around the transition point, there are collective interactions between the atoms, even those far from each other. The collective nature and strong, long-distance correlations around the transition point make describing the phenomena difficult using standard dynamics, which is well suited for differential effects. We have invented highly sophisticated methods of analysis to understand these phase transitions, for example, by averaging the individual behavior of atoms, all the while taking the collective behavior into account. Qualitatively, we understand what is happening quite well. Phase transitions are common in the natural world; they are actually just as natural as a system changing smoothly and continuously. Everyday examples include transitions in bodies between solid, liquid, and gas, but also magnetizing a natural magnet, percolating liquid through a porous solid (filtered coffee!), and superconductivity. Nonetheless, it remains difficult to quantitatively predict the behavior of many elements around the transition — even to predict when the transition is going to occur. For example, it is very hard to predict the temperature of the melting point for a body from what we know of its constituent atoms or to invent a new superconductor material at closer to room temperature.

There are also phase transition situations in fundamental physics, specifically, the moment very soon after the Big Bang when the Higgs mechanism took place and particles acquired their mass. Without mass, elementary particles travel at the speed of light;

it was impossible for them to bond with each other and form any structure at all. With mass supplied by interaction with the Higgs field, their speeds became slower than that of light, which allowed them to bond to each other and gave rise to protons, neutrons, atoms, molecules, all the matter we know — and us. The latter phase transition is hard for us to study today and still presents mysteries, which are undoubtedly connected to the "big questions" we have been talking about from the very start.

We can also bring up turbulence among "nonlinear" phenomena. Under what conditions and at what point does a liquid or gas stop flowing in long, continuous streams (the term is "laminar flow") and give rise to eddies, waves, and foam? How can we usefully describe a turbulent fluid, such as a mountain torrent, since we cannot talk about predicting the position of each eddy at every moment?

In all these cases, despite our sophisticated mathematical methods, our predictive power remains weak. But are we using the correct mathematics?

To attack "big" questions, such as the incompatibility between General Relativity and Quantum Theory, to how great an extent must we challenge our concepts of space and time, and with which mathematical tools? Ever since we identified the problem, we have made numerous attempts — some using traditional methods, others more exotic. We find string theory in the first category. Since the problem arises at tiny distances where fluctuations are huge, let's suppose that the actual physical objects are small vibrating strings. Under certain conditions, we can hope that the vibration patterns combine harmoniously and regulate unwelcome fluctuations.

In the second category, for example, we have "noncommutative geometry," which Alain Connes proposed. Using that approach, we drop our ideas of "points" and "straight lines." We only care about functions that let us find a point in space or on a straight line. Freed from the underlying ideas (point, straight line), we can intentionally choose a type of function that seems better suited to quantum reality. An entire mathematical theory has been constructed to use this new concept; that theory *is not* just an extrapolation of regular mathematics, but something fundamentally different. Unfortunately, this research has not offered a solution to the problem. Still, it remains one of the boldest attempts to challenge commonly used concepts.

Other Ways of Thinking in Physics

Even as audacious a theory as "noncommutative geometry" remains within the scope of our normal way of doing physics: an idea comes from the human mind, is supported by a specific, formalized mathematical tool, and is compared to experimental data. Nothing has changed in this method from Archimedes' discovery of his "principle" in his bathtub (actually, a law of physics rather than a principle) to the invention of the Higgs mechanism. The cycle of "observing/interpreting/imagining the law/comparing it to observations" is essentially similar, even if the theoretical understanding is much vaster and more structured and the experimental means somewhat larger.

Could there truly exist different ways of thinking about how to describe nature and use that description? It isn't impossible. The past few decades have witnessed the appearance of radically different ways of thinking in the sciences. This is the case, for example, with artificial intelligence, the quantum computer, and the study of "emergent" phenomena. We'll return to each of these advances later on, but we can already state that they share some points in common: they are all cases of very nonlinear, very collective phenomena in which there is no longer the possibility of visually following the evolution of each part of the system; only the entire system makes sense. For the moment, we still describe these systems by referring to classical methods: we can examine how an artificial intelligence algorithm works and model the functioning of a quantum computer using a regular, traditional calculator. But can we imagine this type of system becoming independent and grasp its truth without further reference to the "classical" world? That would be a real change in the scientific method.

Education and Conformity

To go beyond current limits, we would obviously like to foster creativity — the ability to think outside the box. But as science gradually advances, the body of knowledge students must gulp down increases. Years of studying physics are barely enough to learn the most classic version of the shared core of general physics and then the state of the art in a very specialized field, with virtually no reference made to the history of ideas, their emergence and evolution, and the doubts and mistakes. We only teach the result, a linear path that is logical, coherent, and roughly chronological. This learning process is prescriptive and does not promote other ways of thinking. While we're on the subject, the system for funding research projects and evaluating scientists by the number of their publications adds to the pressure toward conformity. Going back to education, it so happens that we are currently transforming the habits of the young (and not so young), with the development of smartphones, social networks, and other connections. Certainly, the hyper-rapid circulation of ideas has some threatening aspects. But it could also be used positively alongside education to develop active curiosity and creativity.

Chapter 8

The End of Physics 1: Anthropic Solutions

The Anthropic Principle[1]

The physicist's ideal is reductionist, seeking to "explain" the world by a logical sequence of phenomena from the fewest number of possible of causes. We have to accept that there is a first cause, as otherwise the question of what came before it will arise, and we really cannot imagine that the world came from "absolutely nothing." But even if the universe was created from nothing (using technical terminology, by a quantum fluctuation in the vacuum), who created the laws of physics, the possible types of particles that could exist, etc.? Let's not continue with this ontological question, but rather look at it from the point of view of a physicist today: in the ideal theory, a tiny number of causes lead to

---

[1]From the Greek word, *anthropos*, human, not to be confused with "entropic," which comes from the physical quantity, "entropy."

the entire universe with all its diversity. In today's terms, we would typically see a unique type of fundamental particle of matter and a unique kind of interaction between two or more of these particles. Or perhaps a geometrical entity that would give rise to space and time and whose spontaneous folding would form particles. In any case, as straightforward and pared-down a starting hypothesis as possible. The challenge is to imagine as simple a system for the origin of the world as possible (at least one that seems simple when we describe it in our human terms) that through natural evolution can still generate a universe as complicated as ours appears today.

We have done well traveling along this ideal path, as the Standard Model — with its 12 particles, 4 interactions, and some 30 free parameters — lets us understand how our universe developed and how nuclei, atoms, molecules, gases, solids, liquids, stars, planets, galaxies, and clusters formed — just about everything (except the "big questions," of course). Physicists could be satisfied with that, but the question of "why" continues to plague us. Why these particles? Why these values for fundamental constants? Why this structure for space–time and its interactions?

Certain values for constants are critical for the universe to look the way it does, and in particular, for human beings to exist. Hence the "anthropic" principle, which Brandon Carter articulated around 1970: there is observer bias in our universe simply because we are here to observe it. It's impossible for us to observe a universe in which the laws and constants in physics do not allow humans to exist.

The most commonly cited example of this "fine tuning" of a physical value that our simple existence necessitates is Fred Hoyle's prediction in 1953 that an excited state in a carbon nucleus exists — before this state had been discovered or measured experimentally. Indeed, this excited state is necessary for carbon to be produced in stars at the amount observed in the universe. After the anthropic principle appeared in the 1970s, Hoyle's remarkable deduction was offered as the first "prediction" made from this principle. Carbon is needed to make human beings, thus for observers to exist. It does not actually seem that Hoyle ever connected this abundance of carbon to the presence of human observers. But the argument persists: we, therefore, have many examples of numerical coincidences and precise values that do not seem to be the consequence of any fundamental laws, but are simply necessary for the universe to be the way it is so that we are here to talk about it.

Applying this principle is a bit vague: how do we distinguish between a "natural" quantity and one that's been "fine-tuned," per the needs of the principle? We have now returned to the question of naturalness that we spoke about in Chapter 5.

In any event, it's clear that strictly applying the principle is a slippery slope. After all, why take the trouble to try to understand and explain things if after going back a few reductionist stages, we stop and say, "It's that way because it has to be that way?" We might just as well have stopped at the step before, and one thing leading to another, no longer explain anything at all, content to observe the universe without modeling it, to be the universe's accountants but not its interpreters.

The Multiverse

It's difficult to give the anthropic principle a precise meaning when applying it to a single universe. It becomes a personal choice: certainly, I can observe values compatible with the presence of human beings, and I can stop right there. Or I can also try to understand if the universe is governed by simpler, more general laws from which we can deduce the value of these constants. Choosing between these two attitudes is open-ended: it more or less boils down to trying to come up with statistics from a single sample, since, by definition, we can only observe one universe — our own. But since the 2000s, the idea of applying statistics to universes has become more specific. In fact, several theoretical developments describe the permanent birth and simultaneous presence of a large number of universes. The laws of physics are *a priori* different in each of them. This is what we call the "multiverse" theory.

Superstring theory has provided additional support for the multiverse. As we saw in Chapter 1, the preferred version of the theory envisions space as having many additional dimensions beyond the usual three found in our universe. The extra dimensions are supposed to fold into themselves (called "compactification"), in such a way that they are no longer visible at our scale. There are myriads of possibilities for these compactifications — a little like the way that crumpling a sheet of paper can cause folds, tubes, and rips — and we have no rationale for preferring any one of these possibilities. Despite numerous advances and a good deal of effort, we have not been able to prove that the superstring hypothesis results in our own universe as the only type possible.

Therefore, here is a perfect source for the multiverse! Yet even if we admit that superstring theory is correct (and despite there being no experimental signs of it), no one can know if a multiverse is really necessary or if we simply have not found an explanation that makes our universe the only compactification possible.

Since the definition of a universe is that you cannot go beyond its boundaries if you live in it, the multiverse hypothesis is inherently unverifiable — and thus, unfalsifiable. Therefore, many physicists tend to classify the idea as interesting but unscientific.

As stated somewhat provocatively in Chapter 3, the multiverse theory leads to a form of animism, with a different ontology in each universe.

Between Reductionism and Anthropism

Anthropism does not mean the end of science, nor the end of physics. We might be satisfied to explore mechanisms and relationships, conceding that each time we encounter a slightly difficult question, we will provide another "ad hoc" answer, a new parameter, or a new theoretical patch. This would allow us to make practical use of our new knowledge but would drop the reductionist ideal of understanding in simple terms. You rightly sense all the disappointment this last sentence holds. But without getting to that point, is the border between reductionism and anthropism so distinct? In practice, when we try to represent our universe, we have a choice between a multitude of possible representations too. Our most cherished theories are also derived from the history of human thought, of cultural, sociological, and environmental circumstances. In fact, their contradictions with each other and internally give us ample proof that these are not unique truths we must simply reveal.

Can we imagine another way besides doggedly searching for a miracle solution or anthropic fatalism? Perhaps it's time to consider what our representations of the world are, how we form them, and what we expect of them.

Chapter 9

The End of Physics 2: Algorithms and Artificial Intelligence

Artificial Intelligence in a Few Words

Everyone can attest that computers now do tasks that human brains traditionally performed: classifying, sorting, deducing, suggesting, recognizing, writing a text, composing music, etc. To see the impact of this AI technology on science, it's worth including a detailed example here of how an artificial intelligence algorithm works. Therefore, let's take the standard example of character recognition. Let's first imagine that the letters to be recognized are printed characters from a single font, with each character always written in exactly the same way. It's then clear we can devise a recognition algorithm by placing a character onto a grid and then comparing the list of black and white cells with standard lists of the characteristics that distinguish "A," "B," etc., until we find one that matches. Up to this point, only a logic algorithm is involved — a collection of yes/no tests. If all the squares in the letter to be recognized match those in the reference list, the corresponding letter is chosen. This is also a type of yes/no answer: either there is a solution and the letter matches one of the known letters or we can confirm that the relevant drawing does not. Such a system makes sense, but it is not "intelligent."

An initial improvement would be to be able to tolerate a few flaws. For example, when comparing the standard characters, we might count the number of squares that correspond to each letter on the reference list and then choose the one that has the most matches using a quality criterion — a limitation — of at least X% of squares matching.

A human being who knows how to read can recognize characters with a great deal of variability in the writing, size, and positioning. It would be extremely tiresome to design a pure logic algorithm that takes all these variables into consideration, including all the different types of letters possible in the reference list. What's more, such an algorithm would give no response at all if the character presented did not have enough squares in common with those on the list. Human beings know very well how to accomplish the mix of logic and approximation that lets a character be recognized every time (or almost).

The major change has been our ability to come up with an algorithm that, like a human being, learns from examples. We design a program that inputs the value 0 or 1 for the white and black squares in a drawing placed on a grid, in order to calculate a number between 1 and 26, which will be the position in the alphabet of the letter chosen. For example, if the result of the calculation is between 12.5 and 13.5, we say that the program indicates an "M," the 13th letter in the alphabet. The program carries out the calculation with the help of pre-established functions and a multiplier table (called "coefficients"). In preparation, we train the program by giving it many varied types of characters. We adjust the coefficients in the program so that the rate of recognition within this training sample will be as high as possible, that is, so that the results of the calculation will correspond most often to the correct number in the alphabet for the specific letter. Note that there is nothing complicated about this "training," often shown mysteriously or magically.[1] It's simply the classic problem of adjusting the parameters of a function to get a given result, in this case, the maximum number of correct choices among the examples. Of course, we provide multiple examples of each character, which cover, to a great extent, the possible variability in writing. The coefficients are then fixed and the program is ready to do its work of recognizing the character, like a student who has learned to read from many examples and can then recognize a letter upon seeing it. When we give the program a new character to recognize, it performs the calculation with the designated coefficients and then the result of the calculation indicates which letter is "recognized."

Naturally, the program's ability to perform the task successfully, that is, for it to have a low error rate, completely depends on the types of functions used, their number, and their order. We know of several types (neural networks, decision trees, etc.) that allow programs to discriminate between letters effectively. Their common

---

[1]Most reports on "neural networks" highlight "retro-propagation" as a training method, as if this were an essential, miraculous method. It's nothing of the kind. We can adjust the coefficients via other classic methods of optimization, and incidentally, often much more efficiently.

point is that the basic functions used have to be very "nonlinear" (they move quickly from one value to another). And there need to be many of them in the program to do a good job covering all the possibilities.

Let's take two characters that have points in common, for example, an "I" and a "T." The program has firmly settled on "I" based on small variations in the character, with the calculation resulting in the order in the alphabet of the letter "I" — either a value of 9 or very close to it. Now, let's progressively lengthen the top horizontal line. At first, the program will stay close to 9, which corresponds to "I." Let's lengthen the top line again. At some point, the value will move a little bit away from 9 — 9.2 or 9.3. Then suddenly the result switches to 20, indicating that the program has now chosen the letter "T." When a top line is this new length, it prefers "T" to "I." There's no in between. And there shouldn't be, which means the recognition algorithm must be "nonlinear." If we show a "linear" algorithm a poorly formed "I," with a top bar that is a little too long, thus midway between "I" and "T," it will give a numerical response midway between 9 and 20 (14.5), which in the alphabet corresponds to between "O" and "P," a result that has nothing to do with either "I" or "T."

Once again, the nonlinear, collective nature of the problem — and its solution — is key.

The Example of Experimental Particle Physics

Much of the experimental research in particle physics is done with large accelerators; CERN's LHC, currently in use, is the best known of them. These machines accelerate known particles (protons or heavy nuclei at the LHC) and cause them to collide, thus bringing about very high concentrations of energy in a very small volume and time, which allows us to probe the structure of matter at tiny scales of space and time. Large international collaborative groups of physicists and engineers have installed huge particle detectors where these particle beams meet. For each proton–proton collision at the LHC, for example, a big detector records the direction, energy, and types of all the particles produced by the collision, typically about 50 particles. These "events" (in physicists' jargon) are sorted on the fly. A small fraction of them, the ones we think of interest, are recorded for a more detailed analysis, the goal of which is to better reconstruct the process that gave rise to each event and potentially to discover a new process among all the events or measure an important physical quantity. As is the case with everything that takes place in the microscopic world, these collisions are governed by Quantum Theory, and identical collisions can give rise to different processes. This is even the rule: all possible processes are brought about, with their frequency depending on the process. Thus, we explore the whole range of possibilities by bringing about a large number of proton–proton collisions, for example. The more collisions there are, the more access we will have to rare processes that had not been revealed by prior machines. That was the case for the Higgs boson, discovered with the LHC in 2012.

Even if we have a machine that provides enough energy and collisions to activate the process we are looking for, however, there is still a major problem. These numerous collisions can also generate events that look like the desired events, such as producing the Higgs boson, but are not. They can be everyday processes that appear slightly distorted or measurement errors in the equipment. Even the best instrument has a margin of uncertainty. The problem is thus to distinguish the "signal," the events due to the desired process, with a great amount of "background noise" from everyday processes that simulate it. This is a classic situation in science: all experiments or measurements are subject to background noise and uncertainty, and announcing results consists, above all, of convincing the community that the signal isn't a fluctuation in the background noise or that the measurement is not biased past the specified uncertainty.

Thus, to most effectively reveal a new particle or a new process, the game consists of differentiating as best as possible between signal events and events due to background noise. For each event recorded, physicists know how to use the data the detector provides in order to reconstruct specific portions of it, identifying the known particles and measuring their type, direction, and energy. To retain the most signal events possible and reject the most events caused by background noise, a typical example of the selection criteria could be as follows:

> Choose events that contain at least one particle of one type and of at least a certain energy and another particle of a different type and at least a different, specified energy, their directions separated by an angle of such and such value.

The theoretical calculations and digital simulations will guarantee us that this selection keeps a good fraction of the signal events and rejects a large fraction of background events.

Until around the 1990s, in experiments at accelerators before the LHC, sorting between signals and background noise was done using this type of criteria, based on well-known physical quantities and adjusted by the researchers one by one. The analysis was guided by understanding theory, trying to best maximize the characteristics of signal events predicted by theory. Starting in the 1990s, things began to change in analyses performed on data resulting from experiments at the LEP collider (at CERN) and the Tevatron collider (in Chicago), with the advent of artificial intelligence algorithms to guide the selection. Programs such as these can sort through signal events and background noise much the way they can recognize characters — even those badly written!

Let's imagine that we have reliable simulations of signal events and background noise. We can then train the program to choose from among them by offering it batches of simulated events about which we tell it the type of event (signal or background), adjusting the parameters of the program to provide it with the most correct rate of identification across training batches. Next, to analyze real collisions, we show each event the detector measures to a pre-trained program, which gives us its answer: signal event or background event. The method is ideally suited to these types of analyses in particle physics: the events are independent from each other and there are numerous batches, often millions or billions.

It seems simple, but it took years before the physics community validated the results obtained by this method. There are, of course, objective arguments for this reluctance. It's hard to prove that the simulation is as reliable as we need it to be: like all students, the program only learns well if it's taught properly. Experimental effects — detector or method bias — are harder to understand with a "global" approach such as this than when analyzing via small steps and successive criteria. But the primary sticking point was physicists' psychology: at the time, I often heard a physicist say, "I don't trust a black box." Yet this was just an additional tool for analysis, and we already were using many: geometry, algebra, statistics, etc. Why did this one arouse such opposition?

It strikes me that the real reason is that a method like this takes away a traditional part of a researcher's art. Building an analysis of the data had meant using a physicist's intuition to find a successful series of criteria for discriminating between signal and background. In this version, there wasn't a standard method and we found a path for each new analysis, like a climber finds a new route on an untouched wall.

With artificial intelligence, it's all in the preparation: simulating; pre-processing; and choosing the most appropriate type of sorting program. Then the program carries out the sorting itself and labels each event as probably coming from a signal event or, opposite that, background noise. We evaluate the probability of error in the program by showing it a new simulated batch in which we have deliberately mixed signal and background events together.

After years of attempts and community resistance, artificial intelligence has finally taken hold. Today, all experiments at the LHC use it. Generally, the analysis still consists of a large number of techniques derived from theory and explained by physicists as, "I'm calculating this because that type of background noise will be better isolated, etc." But we have artificial intelligence step in to optimize each stage, identify each particle, or for the overall event, best be able to take advantage of subtle differences between signal and background events. It's true that the simulations have made a lot of progress, making this method more reliable. Today we also understand much better what type of problem benefits from this method and also how to control theoretical and experimental uncertainties despite the "black box" approach.

Returning to the role of the physicist, extrapolating a bit, we can imagine that for whatever analysis, we simply produce simulations of signal events and background noise, which we provide to a good artificial intelligence program for it to learn from. We next just apply this program to real data — and voilà! It then produces the optimum differentiation. At best, there will be no more need even to program a "reconstruction" of the events, nor to perform complicated calculations of the event itself: the black box will take care of everything.

Can We Do Without Theory?

Even more recently, advances in artificial intelligence (we'll call it AI in the following) have allowed us to create algorithms that can identify objects — for example, categories of images that the algorithm was never explicitly designed for. Here's a simple but realistic example. We show this kind of program millions of photos from social networks and the program responds, "I have identified a category of images that has a kind of ball with two little triangles on top of it, small, fine, somewhat horizontal lines a bit below the middle of it, and short, fine, bristly lines around all of it." In other words, all on its own, the program identified the category of "cat photo." It's simply a matter of modeling and has nothing to do with the regular way we model a cat via its zoological, cultural, or other description.

How would a program like that respond if we showed it data from a physics experiment, but without first teaching it about physics?

Beginning in the 1980s, programming researchers have written algorithms that could detect regularities in data from simple experiments and thus succeeded in "rediscovering the law of Ideal Gases or Ohm's law." But it was nonetheless clear that these algorithms depended greatly on human guidance: they discovered what they were designed to discover. Recent advances in AI have made this topic far more interesting. For example, a publication from 2009 made quite a splash.[2] The authors built a simple mechanical system that is well known to physics teachers: a double pendulum. In this system, the first pendulum arm hangs from a fixed pivot; at its lower end, the pendulum has a second pivot that a second arm hangs from. A mass is attached to the lower end of this second arm. When at rest, everything is vertically aligned. We start the system by pushing the mass to the side, away from the vertical, and then we let the system evolve. The movement looks fairly complicated and erratic due to the interactions between the two arms. The authors recorded the positions of the points of the system as it moved and gave that data to an AI program. The goal of the program was to try to detect "regularities," particularly the quantities conserved. The program could use the following basic operations: $+$; $-$; $\times$; and $\div$. For instance, it could "try" a combination such as $v_1 \times v_1$ where v_1 would be the velocity at a specific point in the system. Thus, they had the program "observe" the motion of the double pendulum beginning with different configurations and different velocities. They then asked the program what it saw in the motion.

The result was that the program rediscovered the laws of mechanics: it gave the equations for the conservation of energy and momentum. In principle, this is a remarkable, major experiment. In only a few minutes, a generalist AI program trained only to detect patterns was able to come up with concepts that it took human beings centuries to devise. It did this by bypassing the entire intellectual

---

[2]M. Schmidt and H. Lipson (2009). Distilling free-form natural laws from experimental data, *Science*, 324 (5923): 81–85.

approach that let us extract and refine the concepts of kinetic energy, potential energy, etc., simply by looking at the data.

Today, it's easy to criticize this message and present it as over the top: after all, the experiment was designed with this goal in mind and, once again, the program only had to find what we already knew. To my knowledge, there are no examples as yet of a program pointing out a major concept that deductive science has not already come up with. But this criticism strikes me as short-sighted. It's naive to ask if the question, "Can an algorithm discover a physics theory?" could be isolated from the scientific context of our era. Science, especially physics, is a coherent, constructed system. If we have the intellectual, mathematical, and data-processing means to build AI algorithms and execute them on powerful computers, this implies that we have long understood the elementary laws of physics that these algorithms discover. As a result, it's hard to say whether the AI program's success is just a repetition of what we already know or if the program would have been able to achieve this even without current knowledge, which would only be contingent.

In any event, these experiments and the resulting publications have the merit of asking the question clearly and practically: are we nearing the end of "theoretical physics" in the sense we usually mean? Will we soon be able to bypass the entire sequence we know as the way that physics has historically evolved — defining concepts, then making them more effective via natural selection? By (greatly) extrapolating, we could imagine an entirely different process in the future:

We give a big (huge) AI program all the raw data from all the physics experiments that have already been carried out.

This program, provided with an algorithm and these data, constructs the totality of the currently accepted theory.

In the first place, a good theory can predict the results of new measurements. Thus, to fulfill this function, we ask the program, "What will be the result of my making a new measurement such as this?" The program will give an answer by interpolating/extrapolating from the measurements that it already knows, exactly as our character recognition program did.

Next, a good theory needs to be productive, for example, to offer practical applications for human beings. It's also possible to ask the program — which knows all human systems, objects, and facilities — whether it can find any refinements and even if it can come up with new applications. Thanks to its infinite ability to calculate, the program can always test all possible systems and choose the ones that will have a chance of working. No doubt it will suggest new, useful ones.

A good theory should also be fertile for ideas. But since there's no longer any need for ideas, why search for new ones? Well, if human beings really insist on using their old blueprint for attaining understanding, the program could certainly suggest new patterns that it will have detected in the data — new relationships, new equations. True progress would, therefore, only be in the program itself. There would be nothing to preclude us from asking it, "Do you see anything in the data that could help make you better?" and even, "What new experiments do you need in order to become better?" These new experiments, suggested by the algorithm itself, would then be the basis of the scientific program.

I feel you shudder. Human beings seem to play a very minor role at this stage.

We are very far from this and nobody knows if we will actually go in that direction. With each publication, raised voices remind us of the central role of human imagination and creativity in scientific progress. I agree; this is clear and today we can still easily reassure ourselves about the preeminence of the human mind. But such research is valuable, as it thoroughly questions the scientific method. The abovementioned voices are actually captivated by great intellectual creations, such as mechanics, electricity, and classic thermodynamics, particularly when a theory is apparently the work of an individual brain, as in the breathtaking example of General Relativity, which Einstein discovered. This fascination bestows a transcendent, even spiritual character upon the creations. I think this way of seeing wrongly minimizes the mechanistic aspect of theories: a theory is merely identifying concepts and patterns, and we validate it as soon as extrapolating it onto nearby territory is shown to be accurate.

Even if replacing all theory is not part of the future, the example of "all AI" should at least encourage us to be somewhat humble.

Chapter 10

The End of Physics 3: Disenchantment

The Original Sin: The Atomic Bomb and Its Use in 1945

On August 6, 1945, a flash of lightning lit up the sky above the Japanese city of Hiroshima. That single flash, the heat of its radiation, and the firestorm it created reduced the entire city to rubble and killed around half its residents, between 100,000 and 150,000 people. Three days later, the same flash erased the city of Nagasaki, killing about 70,000 people. Sadly, this wasn't the first time that bombing razed a city and killed tens of thousands of inhabitants, but the world learned afterward that a single bomb had caused each of these massacres. Everyone quickly understood that "atomic" (rather, "nuclear") weapons would determine relationships between major nations for a long time to come. Above all, the entire planet learned that this enormously powerful weapon resulted from 20th century physics implemented in great secrecy by the U.S. We linked the weapon to the personality of one man, Einstein, and we were simultaneously terrified and amazed by the juxtaposed images of the devastated Japanese cities and the portrait of the waggish scientist. As so often is the case, this shortcut involved both truth and fiction, beginning with the fact that most famous photos showed Einstein at over 65 years of age even though he was only 26 when he discovered the theory of Special Relativity, which unveiled nuclear energy.

Of course, the world was already aware that science could simultaneously be the vehicle for the greatest progress — health, mechanization, transportation, communications — and the greatest destruction — conventional bombs, explosive shells, poison gas, etc. But this time, the event was unanticipated, as it had been secret — a single event, two bombs, had stopped a war (at least that was how it was presented at the time) — and inconceivable. Its power did not correspond to anything then in use, compared to conventional explosives for warfare and also public works, and close in principle to regular fires.

The suffering of the citizens of Hiroshima and Nagasaki profoundly affected how physics was viewed, first and foremost, nuclear physics but also all the sciences. It showed scientists' inordinate power and responsibility. And it also revealed that science could act in a way that was disconnected from daily life, evolving in an intellectual, experimental world cut off from people's needs and fears. Military secrecy was only the most extreme form of the gulf between science and people. Even though most of the Allied countries at the time approved, using these terrible weapons irreparably harmed people's trust in science. Another perspective is that the gulf already existed before then, but the extreme brutality of nuclear bombing revealed it fully to a far greater extent than previous horrors had done.

Overcoming Mistakes: The Example of Global Warming

Today, the fact that science can lead to dangerous mistakes as much as real progress is obvious to everyone, and rightly so. Blindly believing in technological progress without asking questions about responsibility or ethics — as we did until the 1970s — could only have led to a failure of science and society. In response, starting in the 2000s, we have arrived at a widespread mistrust of science, mixed together with mistrust of political systems and, to make a long story short, other human beings.

Yet scientists have also shown themselves able to realize their mistakes and suggest corrections. Studying global warming strikes me as a meaningful, important first example. In fact, scientists were the first to identify the human causes of warming and warn of potential consequences. Without falling back into blissful optimism about scientific actors and habits, we can rejoice that over the past half century, scientists' roles and responsibilities have been the opposite of before where this subject is concerned. Simplifying things a bit, today it is scientists who are warning and encouraging caution so as to protect future generations, whereas most of the world's population egotistically resists changing its lifestyle. Luckily, another part of that same population already supports scientists, and we can see a path toward reconciling and recovering science.

Chapter 11

A New Age of Enlightenment?

The Age of Enlightenment

I s it a French bias to marvel at the extraordinary cultural period that heralded the arrival of Enlightenment philosophy?

The Enlightenment offered a political evolution: the rights of man, the abolition of privilege, and the advent of democratic societies. But this evolution took place alongside a noble, universalist scientific vision.

My favorite example of this revolutionary, selfless process is the invention and implementation of a universal measurement for distance: the meter. Before the French Revolution, all measurements of distance, volume, weight, etc., were specific to the country and even the region. Speculators and those who knew the system best had a field day with the confusion this created at the expense of their poorest customers, whom they could deceive about an amount of gold, fabric, or grain. During the Revolution, the National Constituent Assembly — through Talleyrand's advocacy and inspired by scientists and philosophers in the Academy of Sciences such as Borda, Lagrange, Monge, Condorcet, and Lavoisier[1] — suggested establishing a universal system for units of measurements that all human beings anywhere on Earth could understand. For length, and thus also surfaces and volume, the basic unit would become the meter, defined as one forty-millionth of the Earth's meridian.

While preliminary measurements of the meridian existed, they were difficult to standardize. So the Assembly charged two astronomers, Delambre and Méchain, with measuring the distance and the curvature of the Earth between Dunkirk and Barcelona. To do so, they had to go between the two cities using telescopic sights to measure from steeple to steeple, farm to castle, and hill to mountaintop in a France that was at the height of the Revolution and even, at the end of their travels, at war with Spain. The venture would take seven years, from 1791 to 1798. The account is an amazing, captivating page in the history of science. It's moving, too, for Méchain, having been wounded, imprisoned, and then freed, died of yellow fever in 1804 in Barcelona, where he had returned to try to find the source of a tiny discrepancy in the measurements he could not explain.

These two eminent scientists — peaceful astronomers already celebrated in their community — did not hesitate to hurl themselves onto roads and contend with danger, for the sole purpose of

---

[1]Despite his vast, noble project, the unfortunate Lavoisier was guillotined in 1794 because he was a tax collector. At his mockery of a trial, the president of the revolutionary tribunal declared, "The Republic does not need scientists or chemists…"

giving humanity a reliable, neutral, honest unit of measurement. The assembly of elected representatives of the French people gauged this project to be so important that they decided to do it, supported it, and then insisted on it despite the inevitable resistance.

Another major Enlightenment project was writing and publishing the Encyclopedia, accomplished by Diderot and his friends. That project, too, moves us with its nobility and the perseverance it took. The goal of this first encyclopedia was to put all the knowledge of the time in everyone's reach. Including geography and history allowed each person to find their place in the world, instead of only knowing that of their birth. With technical knowledge about agriculture, different crafts, etc., they could also become self-sufficient. Finally, everyone could now think about subjects that up to then had been regulated by church and state authority: religion and philosophy.

Today, the internet has also revolutionized access to a rich, virtually limitless source of information for the greatest number of people. The original motivations for inventing the World Wide Web were not all that far from those of the encyclopedists: the web was invented at CERN to allow researchers from each country and every culture to collaborate efficiently, as researchers are the connective tissue of this international research center. Unlike the Encyclopedia, which limited the role of the public to reading it, thus passive, the internet is now a participatory network where all can contribute, for better or for worse. A single example of "for better" suffices: Wikipedia has become the primary encyclopedia of first resort for the public, and, let's face it, for experts, too. I am amazed by how civilized the contributors are and by their sense of responsibility. The pages for building and revising articles are often fascinating, with their polite, rational debate. As for "the worst...," I do not need to describe them here: you know them all too well.

Current Successes in Physics and Some Future Paths for Revolution

A large part of this book has been devoted to showing the crises and problems in today's physics. We should not, however, let that make us forget the extraordinary successes in science, and this field, especially. We only find ourselves facing fundamental questions, the "big" questions, because recent advances have allowed us to explore the largest and smallest dimensions of our universe experimentally and theoretically. Once again, it isn't shocking that in the course of such immense, swift progress, there may be moments of crisis or impasses that will require a bit more time to resolve.

I would like to mention two paths that seem relevant to me here, one theoretical, "emergent gravity," and the other practical, the "quantum computer." A word of warning, though: I am not about to claim that either of these two paths is currently in a position to solve all the abovementioned problems in their respective fields of quantum gravity and the information revolution. But it strikes me that these two paths are evidence of the *spirit* of revolutions to come in physics. We shall see how after briefly discussing them.

The phrase, "emergent gravity," refers to a specific work (one publication and others following it) by Verlinde, a controversial study in the theoretical community. But the proposal fits in with more general thinking along the same lines. These ideas stem from theoretical studies about black holes, for example, when we try to understand where the information contained in objects that fall into a black hole goes. This question turns out to be fruitful, as it requires an unusual point of view that lies between the vision in General Relativity (classical and geometrical) and that of Quantum

Theory, which is connected more to subtle and microscopic properties of objects. According to this school of thought, information is the primordial essence, before matter and gravity. Gravity would have come into being as the result of laws that are more fundamental and that govern the presence, density, and exchanges of information. This is the sense in which gravity and matter would be "emergent."

It's striking that a theoretical framework with information as its essential element would appear today, at a time when information in the common sense of the word is also being highly developed and at the center of society's concerns. Either this is a cognitive bias that borders on naivete — in a society where everything revolves around information, theoretical physics superficially gets with the program — or else the path is correct and the physics of the future will actually be based on the concept of information. This future physics would naturally also be in sync with society regarding the information revolution. The truth is undoubtedly somewhere between these two possibilities.

The "quantum computer" may be another aspect of the information revolution that might come from fundamental physics. In this physicist's dream, we would use the uncanny properties of quantum objects interacting collectively at a distance from each other to solve complex problems. Our understanding of a coherent group of quantum objects is only beginning and there are numerous obstacles. In particular, the slightest interaction of a group like this with its "classic" environment can destroy the coherence that was obtained with such difficulty. Nonetheless, it seems probable that researchers will overcome these obstacles and figure out the quantum computer one day, perhaps even in the near future.

It's hard to imagine the true power of such a computer and even more to evaluate the change in mindset that would accompany its arrival. Would human beings learn to formulate their questions in a highly parallel manner, rather than "linearly" and inductively as is done today? Still, this research fits in beautifully with the perspective of nonlinear, emerging, comprehensive thinking.

An Age of Enlightenment Today?

Asking this question may seem to show naive optimism that is out of touch with today's world, a world that looks unpredictable and violent to us. Countries that have long been on the path toward liberty and freedom of thought, universal education for boys and girls, improving living standards, and increasing knowledge, are being tempted by fits of populism and intolerance or else they have already succumbed. Only a few voices remember that the human condition continues to improve worldwide (Pinker). And yet, this material improvement comes at the cost of environmental disaster. The dream of an internet that would be a powerful tool for developing culture, citizens, and relationships has been hijacked and is heavily exploited for the profit of a few sprawling firms. In short, the picture is bleak.

To the bleak picture that the world news paints for them — as well as for each of us — physicists add the crises in the sciences and at the heart of their own field, the crises in physics we spoke about earlier.

But if we do not give in to the catastrophic thinking around us, we can also think about which ingredients will help us escape the above crises. We just have to go along with developments instead of fighting them.

What would an Age of Enlightenment be like in the era of AI? Let's return to the heart of physics, to the "big questions" I've mentioned. We're still mesmerized by major individual achievements such as those of Einstein. Yet even though we know that physics, like all the other sciences, has been shaped by multiple contributions, we still look for "Eureka" answers to our questions. Perhaps that will be the case: perhaps an attractive theory will come to replace Quantum Theory and Relativity tomorrow. But maybe not. Perhaps, instead, an algorithm will interpret our experiments and suggest new ones. Or maybe the future will lie somewhere between these two opposites: an algorithm will help us interpret data and design new experiments, and we will contribute — all of us rather than a few geniuses. We might even imagine that our future representations of the world will require a harmonious blend of human and artificial intelligence. Some parts of these representations will be supported by concepts in the traditional sense, others by AI units, the "black boxes" that will handle large quantities of complex data for us — "augmented" intelligence, as it were.

The gap separating people — non-experts — from scientific knowledge has increased. But using technology and especially information technology has become completely universal. Most people, even those who are underprivileged, have internet access or they will in the near future. They also have access to extremely powerful data-processing tools. A phone today has the same data-processing ability as an entire computer center in the 1980s. And people easily take ownership of what the phone can do: they create web pages and videos, compose music, etc. The possibility of creating and distributing texts, music, and videos of oneself, without needing to call upon art or publishing institutions, has undeniably liberated and encouraged a good deal of creativity.

There's a bad side to this proliferation, to be sure: the flood of useless information; the explosion of social networks and their shallow communications; rumors and fake news; the risk to individual freedom; and most of all, the intense commercialization. But we can also imagine new growth from all our use of scientific progress. Extending the process a bit, we can imagine a type of participatory science much like the new artistic participation: even if I don't understand the complicated equations in the Standard Model and the Higgs mechanism, I can still be involved in the research applying the same algorithm that helped me identify photos of my friends on social networks. I can — and this is already the case, incidentally — make my computer or phone available to researchers at night to provide them with computational capacity.

If even scientists concede that algorithms should seamlessly complete theories, then the burden of difficult concepts, repulsive mathematics, and years of abstract studies can be reduced. Not that they would be devalued; quite the opposite! But everyone would be able to access knowledge and representations of the world at their own level, with the right personalized "mix" of pure concepts and algorithms.

Chapter 12

Toward a Rationalist Spirituality

Science and Aesthetics

The relationship between science — and, for my part, physics — and aesthetics is complicated. On the positive side, there are many initiatives for encounters between researchers and artists, and it's always fascinating to participate in them.[1] We understand each other well: we're on the same quest to represent the world, to find imagery that "speaks" to the human spirit. We also share a passion for discovering, creating, and personally interpreting the world. During these encounters, the public is always present, thrilled to see these two visions in conversation. At least, the public that is educated and curious. I'm well aware this is not a vast audience in terms of society.

---

[1]One of my first contacts with the arts community as a physicist was by participating in the exhibition, "Mathematics: A Beautiful Elsewhere," at the *Fondation Cartier*, Paris, in 2011, an adventure that was as pleasant as it was rewarding. Thank you to Michel Cassé for having enticed me to go!

The less positive side is that beyond these events that are organized before a friendly audience, the general public continues to view science — and especially a "hard" science such as physics — as hostile, daunting, and even inhuman. Engineers who simply talk about what they know find themselves rejected because "their world completely lacks poetry." As I said above, researchers under pressure from the media often turn to beautiful images to make their subject engaging, although that's often just a trick. But we need to convince the public and the mainstream media that the scientific vision of the world is also an attractive one and to cease setting the two types of representation in opposition to each other. As I tried to show in my previous book, equations take nothing away from the beauty of the world. Quite the opposite, they emphasize it: isn't it miraculous how the law of refraction acting on the rays of the sun in raindrops shows us a magnificent rainbow?

Science and Spirituality

The chasm between the scientific representation of the world and how it is perceived by the public is so great that for most people today, science and spirituality are completely contradictory. On the one hand, the need for spirituality has never been greater, with our highly materialistic lives, distanced from nature and greatly restricted in terms of time, money, and shared rules. On the other hand, there is science, which constantly provides knowledge, tools, and comfort. But it absolutely does not fill the spiritual void.

When we engage with audiences at conferences for the general public, meetings, science cafes, and lab open houses, we can see that science always makes people dream: we encounter people who are eager to learn about the latest discovery in the world of particles or the universe. Of course, this is mostly a select audience. And even for this audience, the information that scientists bring satisfies their curiosity, but most often it does not really change their worldview. Nothing could be more normal, as the communication usually is one way. The public is passive and does not participate. The scientific discussion remains a pretty tale, but it doesn't become a part of real life.

A small participatory moment comes with the questions. Incidentally, my favorite questions are not those from experienced amateurs, whose questions come close to those by professionals at a conference. No, my favorites are the most spontaneous questions, the ones that seem the most naive. At first, I was thrown when I had spoken about the Big Bang and someone in the audience asked me about the existence of God. What to answer? How even to receive that question? We scientists must not lose sight that deep down, it is the same quest for meaning. That question from the public is perfectly legitimate after my talk, even if I can't provide an answer (but religion doesn't provide more of one either).

Integrating Artificial Intelligence

We need to face facts: within a few decades, all human beings, or almost all, will be equipped with an AI tool in one form or another. This tool will constantly communicate with all the others and will have an unlimited ability to process data for all practical purposes. *All* information from *all* humans will be accessible to *all* (unless we let a few large companies retain control of it — but that's a political question). Things will *change.* As with all large changes in the means of communicating human thought — writing, printing — this will cause a profound reworking of our representations of the world. We can expect the worst as well as the best. This change is the result of science and technology: it would be a shame if practitioners in these two fields didn't use them to help resolve their crises. In any case, there can be no question of staying on the sidelines, refusing to use algorithms for our theories, or putting up digital resistance in order to preserve a false independence of thought or even just fictional freedom. We must take advantage of AI to shake up our paradigms.

Total Mindfulness, Total Awareness

In our overworked Western societies, meditation is in fashion. Needless to say, we shouldn't be too naive; this fashion is for the affluent, where the need for meaning and well-being takes precedence over essential needs of food and physical safety. Nonetheless, let's think about this so-called trend and the apparent paradox it represents.

In a highly materialistic, well-equipped, fully optimized society, this trend is about connecting with the present moment — being fully conscious of the world and our place in it. The paradox is that many view the practice of meditation or one of its many equivalents as a way to escape from an overly materialistic notion of the world: to *reconnect* to the present moment, we would need to *disconnect* from a world that seems too fast, too complicated, and too frenetic, a world that we more or less consciously have lumped together with a world that is too technological and too scientific. Yet this approach of reconnecting to the real and the present is, or should be, precisely that used in science. Physics — at least the way I wish to do it — involves exactly the same approach: searching for a current representation of the world that is based on those of the past without being trapped by it and without presupposing those of the future, a representation whose limitations we are fully aware of — as we saw earlier in this book; also searching for a better representation and the best description of our place in the universe, not necessarily in terms of efficiency, but rather in terms of its consistency, naturalness, and, at the end of the day, beauty.

Toward a Rationalist Spirituality

Once we state the "need for meaning" in these terms, why not imagine that the "total mindfulness" sought for by meditation and its equivalents recognizes the importance of a type of "total awareness?" From my window near CERN — the center where I carry out my research, a center often called the temple of modern physics — I can see Mont Blanc. Often, in good weather, I contemplate this singular summit that dominates the landscape and like everyone else, I can't help finding the view simultaneously beautiful and profoundly moving. And I think about all the laws of physics I know that led to my good fortune of being a tiny, enraptured human looking out at Mont Blanc:

> The Big Bang — the creation of primordial matter, which I do not know about, and the fundamental form of interaction, gravity. The appearance of the elementary particles as I know them today: quarks, leptons, neutrinos, in the form of a massless, shapeless, infinitely turbulent gas. The annihilation of all antiparticles and their corresponding particles, except for a tiny fraction that improbably

survives the cataclysm. The abrupt arrival of the Brout–Englert–Higgs mechanism and the appearance of mass for all these particles, which, as a result, are moving quickly but at different speeds, like a film that is at first fast-forwarding crazily and then suddenly is slowed down to 24 frames/second, so that the movements of pedestrians and cars makes sense. Briefly passing through a chaotic stage of highly intense exchanges — quark-gluon plasma — still more exchanges between photons and charged particles, and then another abrupt transition in which each proton captures an electron, much like a dance where the partners stop changing and suddenly begin to dance as couples: atoms. Time slows. Atoms randomly spread out via large forces created by dark matter, which I know almost nothing about. Then the atoms recombine, falling into groups one on top of the other, and then into increasingly massive clusters. Suddenly, some of the highly condensed clusters light up spontaneously like will-o'-the-wisps, and voilà! The stars begin to shine. This one or that one will shine for a few billion years and then suddenly collapse in an incredibly violent explosion that can create all the chemical elements I know of. The stardust again comes together from what's left, revolving around an average star: our sun. Iron atoms condense into dust and then chunks, forming the core of a planet that becomes our Earth. Nearby, nuclei of silicon and oxygen combine to form crystals, which stick together and around the iron core, becoming little pebbles here, then big rocks, and then — after some geological upheavals that, with my apologies, I'll skip over the details of out of ignorance — here, finally, are the huge faces of granite of the Mont Blanc massif that tourists delight in seeing and climbers thrill to touch. To this miracle, a little water is added, creating the snow that covers the rock and an atmosphere that lets me breathe. And there it all is! Everything — as far as the landscape is concerned… But all these combinations of quarks, electrons, and atoms also provide the building blocks — carbon, hydrogen, nitrogen, etc. — to create the strange phenomenon of life, the extraordinary DNA molecule, the kingdom of life, and particularly — what luck! — me, as I contemplate the landscape.

Whew!

With such a tale and faced with an origin story like this, do we really need more *meta*-physics? When it comes down to it, yes, of course. Even with all the knowledge in physics, the field gives no answer to the question of ontology: why all that took place. But you must admit that it already supplies us with a fascinating, prolific, consistent story for those who are looking to find their place in the universe. Everyone can supplement this story in their own way. Everyone can question it; everyone can try to have their own version shared with others. Everyone can have the pleasure of discovering and making another person's vision their own. Nothing is etched in stone. This revelation does not come from an ancient miracle but from humanity's slow work of continuously observing nature, communicated from one person to the next.

Isn't that *enough,* at least for everyday life?

In short, could we finally contemplate a *rationalist spirituality?*

Chapter 13

Believing in Physics Despite the Crises (or Because of Them)?

Throughout this book, I have shown you the multiple profound crises in today's "fundamental" physics without minimizing them. There is the societal crisis, with the most ridiculous beliefs such as a flat Earth, the mistrust of technology, which physics plays a central role in due to the history of nuclear weapons, and the questions about the concept of progress — despite all its benefits. There is the crisis in physics itself, which has faced internal problems for quite a long time now, with the conflict between Quantum Theory and General Relativity, and with the difficulty in simply accounting for observations in terms of dark matter and dark energy. Finally, there is the crisis in how we represent the world, with the advent of artificial intelligence, and with, at the very least, serious questions about our usual way of thinking about the world by looking at it and building more or less abstract models that human minds can still imagine from A to Z.

We would have reason to feel seriously shaken, even worried, faced with this apparent accumulation of crises. But as I've described them, I've also tried to show how all this questioning could be a source of renewal and reorganization.

As for the internal crisis, one that we could almost call "technical," we should look on the bright side: it's only a crisis because our theoretical understanding and means of observation have evolved considerably in the last century. So it isn't surprising that we now need some time to digest, reflect, and put our ideas in order. As I said in Chapter 2, it's possible that the internal crisis in physics will be resolved conventionally with a new solution, maybe even that a new star researcher — in other words, a new Einstein — will propose it. Experts would welcome this, as they would feel reassured by seeing the usual pattern continue. But that would not resolve the societal crisis; there's a good chance that this new solution would call for even more abstract, more complicated mathematics than what we use today, making it hard for the public to feel connected, once again, and sending them back to their apparent ignorance. Nor would this integrate artificial intelligence any more than it is now, even though AI will become an increasing part of our lives and how we view the everyday world. Such a solution would highlight even more the gulf with the public — who increasingly think with their phones or an equivalent portable AI — while insiders would continue to use purely intellectual processes.

We might instead find it more interesting and even wiser to try to take advantage of the conjunction of these three crises to consider overhauling physics. In all humility, I do not have a "silver bullet," nor do I think one exists. I've insisted to some extent on the role that I think AI must take in solving the problem and overcoming these crises, but that doesn't necessarily mean I consider AI the only "silver bullet." At least, let's not reject integrating it into our way of thinking, teaching, and talking about science. In other words, let's not relegate it *only* to the world of commerce and daily news.

We should also think about how we teach physics, which to my mind is far too theoretical in the bad sense of the word, particularly in France. Physics education in elementary or high school aims to get children to understand complicated theoretical concepts. This is well intentioned, as the concepts of Newtonian mechanics, classical electricity, Quantum Theory and Relativity have changed our worldview. We try, therefore, to get students to understand the solar system or the atom. But because students lack the mathematical background underpinning the real theories, they reach an abstract but inoperative description that is disconnected from observation and reality. At worst, students become able to answer a question about gravity by parroting a paragraph from the course, even though they're convinced the Earth is flat. This type of cognitive dissonance is typical in our societies, far beyond physics, for example, in topics about health, too. Let's teach everyday physics; let's observe nature and the human artifacts around us. Let's use apps as needed to assimilate our observations and forums to discuss them.

As for researchers, let's try to collect "good" measurements that we will attempt to integrate into an increasingly comprehensive model. In terms of theory, let's not let ourselves be dragged down the slope of "building models," in which we ceaselessly repeat the same schema in principle, changing only a parameter, a symmetry group, or a unification coupling. We've done this for a half century with a certain amount of success — the Standard Model — but it's time to move on to something else. Let's develop the concepts of spontaneous organization, of emergence, which we see the beginnings of today.

I like to think that these crises carry the elements for their own solution within themselves. As you will have understood, I am primarily worried about the crisis between society and science, as the chasm has widened, and particularly with elemental physics, because I tend to feel guilty for its use in Hiroshima and Nagasaki, even if this guilt is completely anachronistic: I hadn't even been born in 1945. But that tragic event grows distant; it's time to rebuild the relationship.

Index